D1083109

QUALITY BY DESIGN FOR BIOPHARMACEUTICALS

To our family:
Bhawana, Payal, Parul, and Jyoti

CONTENTS

9 APPLICATION OF QUALITY BY DESIGN AND RISK ASSESSMENT PRINCIPLES FOR THE DEVELOPMENT OF FORMULATION DESIGN SPACE   161

Kingman Ng and Natarajan Rajagopalan

10 APPLICATION OF QbD PRINCIPLES TO BIOLOGICS PRODUCT: FORMULATION AND PROCESS DEVELOPMENT   175

Satish K. Singh, Carol F. Kirchhoff, and Amit Banerjee

11 QbD FOR RAW MATERIALS   193

Maureen Lanan

FOREWORD

These are truly exciting times to be involved in the development of biopharmaceutical products. As the research community expands our understanding of the biological basis of health and disease, those who turn this knowledge into medical treatments are providing safer and more effective health care options. Over the relatively short history of biologically derived drugs, this trend is clearly apparent. The first biological products to be developed were natural products such as antisera and hormones purified directly from animal tissues. The development of hybridoma technology in the 1980s allowed the preparation of monoclonal antibody products and significantly reduced the structural variability characteristic of polyclonal antibody products. The molecular purity of these products allowed them to be extremely well characterized and also led to a much better understanding of the biological activities of their structural features. Subsequently, the application of recombinant DNA technology to biopharmaceutical development has allowed manufactures to design proteins with specific structural and functional characteristics that give them desired beneficial therapeutic properties and reduce their potential adverse reactions.

These changes in product development and expression system technology have driven, and relied upon, parallel advances in the manufacturing sciences. Biopharmaceutical manufacturers have always been faced with the challenges of finding ways to make living systems produce proteins with desired characteristics, purifying them from complex mixtures with economically feasible yields, and formulating them in to stable, medically useful products. These challenges are compounded by the variabilities in raw material quality, equipment components, environment within the manufacturing facility, and capabilities of operators. As those who have struggled with these issues know so well, the quality of biological products depends to a large extent on the design and control of the manufacturing process.

It is crucial to public health that the drugs upon which we depend are safe, efficacious, and of consistent high quality. Safety and efficacy determinations are based on toxicological data, clinical study results, and postmarketing evaluation-based performance. Because the quality of a drug product can have a major impact on its clinical performance, successful drug development and manufacturing must focus on quality. In this regard, the concept of quality is twofold. One aspect of product quality is the design of the drug itself as defined by specification of the characteristics it needs to have to treat a disease. This includes the structure of the pharmaceutically active molecule itself, as well as the formulation and delivery system that allow the therapeutic to reach its target. The other aspect of quality is the consistency with which the units of a batch or a lot of product meet the desired specifications. As was alluded to earlier, within-batch variability and batch-to-batch variability depend, to a large extent, on the quality of the raw materials and the design of the manufacturing process and its control systems. Incorporation of these two aspects of quality into product and process development is the essence of quality by design.

To realize the full benefits of quality by design, one must develop a thorough understanding of the interrelationship between the attributes of the input materials, the process parameters, and the characteristics of the attribute of the input materials, the process parameters, and the characteristics of the resultant products. With this information in hand, it is possible to manufacture with a very high degree of assurance that each unit of product will have the desired quality. Of particular note in this regard is the quality control system known as process analytical technology (PAT) that has been applied with great success to manufacturing operations outside of the pharmaceutical industry. A cornerstone of PAT is the use of rapid analytical techniques and process control systems to monitor and control product quality during manufacturing. In 2004, the FDA published guidance for industry on PAT[1] to encourage the development and implementation of the agency's "Product Quality for the 21st Century Initiative," as PAT can provide the assurance of quality in a flexible manufacturing environment conducive to streamlined implementation of innovative technologies. The use of correlated metrics of quality, such as bioreactor conditions, within the process control system is quite familiar to biopharmaceutical manufacturers. However, future strides in rapid, real-time analytical technologies promise to make direct control of product quality during manufacturing a reality and open the door to efficiencies such as continuous processing and real-time release.

As biotechnology moves ahead, the concepts of William Edwards Deming and others that quality must be built into products will continue to be applied to the design of novel products and dose delivery systems as well as to the design and engineering of more effective and reliable manufacturing methods. Technological advances in this field will undoubtedly occur in an evolutionary manner, with successful systems serving as the foundation of even more valuable systems. However, this steady progression will, nearly as surely, be punctuated by revolutionary discoveries of magnitudes equal to hybridoma technologies that introduced monoclonal antibody production or the polymerase chain reaction that has made genetic engineering a relatively facile process. To ensure that we have the safest and most efficacious medications to treat today's disease, and those of tomorrow, we must not only continue developing innovative products and technologies but also take them to the manufacturing plant and the marketplace as quickly as possible. The sharing of ideas, information, and experience through books such as this is essential to the success of this endeavor.

Keith O. Webber

Deputy Director, Office of Pharmaceutical Safety,
Food and Drug Administration

---

[1]FDA Guidance for Industry: PAT—A Framework for Innovative Pharmaceutical Development, Manufacturing, and Quality Assurance (September 2004).

PREFACE

Quality by Design (QbD) is receiving a lot of attention in both the traditional pharmaceutical and biopharmaceutical industries subsequent to the FDA published "Guidance for Industry: Q8 Pharmaceutical Development" in May 2006. Key challenges in successfully implementing QbD are requirements of a thorough understanding of the product and the process. This knowledge base must include understanding the variability in raw materials, the relationship between the process and the critical quality attributes (CQAs) of the product, and finally relationship between the CQA and the clinical properties of the product. This book presents chapters from leading authorities on a variety of topics that are pertinent to understanding and successfully implementing QbD.

Chapter 2 by Kozlowski and Swann provides a summary of QbD and related regulatory initiatives. Approaches to relevant product quality attributes and biotechnology manufacturing have been discussed along with some thoughts on future directions for biotechnology products.

Chapter 3 by Narum presents a case study where QbD principles have been applied to make significant improvements in the capacity of recombinant expression systems to produce malarial proteins by introducing synthetic genes for *Pichia pastoris* as well as *Escherichia coli*. It is shown that the use of synthetic genes not only makes possible the expression of a particular protein but also allows the gene designer to make appropriate modifications to increase product quantity and quality.

In Chapter 4, Schenerman et al. present a risk assessment approach for the determination of the likelihood and extent of an impact of a CQA on either safety or efficacy. Examples are used to illustrate how nonclinical data and clinical experience can be used to define the appropriate risk category for each product quality attribute. The attribute classifications then serve as a rationale for product testing proposals, associated specifications, and process controls that ensure minimal risk to product quality.

Chapter 5 contributed by Hoek et al. presents a case study involving a cell culture step. All operational parameters were examined using a risk analysis tool, failure mode and effects analysis (FMEA). The prioritized parameters were examined through studies planned using design of experiments (DOEs) approach. Qualified scale-down models were used for these studies. The results were analyzed to create a multivariate model that can predict variability in performance parameters within the "design space" examined in the studies. The final outcome of the effort was identification of critical and key operational parameters that impact the product quality attributes and/or process consistency, respectively, along with their acceptable ranges that together define the design space. Chapters 6 and 7 define approaches to establishing design space for a filtration and chromatography unit operation, respectively.

Sofer and Carter present a strategy in Chapter 8 for applying QbD principles for virus clearance. It is concluded that implementation of the proposed strategy will require an extended and coordinated effort, primarily by manufacturers and regulators. The mutual investment in moving to a QbD approach holds promise of better understood, and therefore better controlled, unit operations. The QbD design space describes a full range of manufacturing conditions within which changes may be made with relative ease and modest regulatory oversight, freeing both manufacturers and regulators' limited resources. Intrinsically, enhanced process control and process understanding represents a benefit to the patient population.

Chapter 9 by Ng and Rajagopalan presents the different considerations to remember while designing a formulation process. Some of the key steps include identification of target commercial drug product profile; preformulation and forced degradation studies to characterize molecular stability properties, impact of formulation variables, and other factors; preliminary stability risk assessment with emphasis on direct impact on the activity based on preformulation and forced degradation studies results; initial formulation risk assessment to establish the cause–effect relationship of different factors and solution formulation stability via Ishikawa (Fishbone) diagram; multivariate DOE studies to optimize the formulation composition and define a robust design space to meet the expected shelf life of 24 months at 5°C; establishing formulation design space based on DOE results and stability properties projections; and finally selection of commercial solution formulation based on design space, molecule knowledge, and risk assessment.

In Chapter 10, Singh et al. present case studies illustrating a systematic work process for application of risk-based approaches to formulation development for biologics.

Lannan addresses the application of multivariate data analysis (MVDA) to analysis of raw materials in Chapter 11.

Chapter 12 by Molony and Undey provides a review of various PAT tools and applications for the biopharmaceutical industry. Finally, Chapter 13 by Low and Phillips provides the background for PAT and also how it relates to QbD.

Anurag S. Rathore

Rohin Mhatre

Thousand Oaks, California
Cambridge, Massachusetts
March 2009

PREFACE TO *THE WILEY SERIES ON BIOTECHNOLOGY AND RELATED TOPICS*

Significant advancements in the fields of biology, chemistry, and related disciplines have led to a barrage of major accomplishments in the field of biotechnology. The *Wiley Series on Biotechnology and Bioengineering* focuses on showcasing these advances in the form of timely, cutting-edge textbooks and reference books that provide a thorough treatment of each respective topic.

Topics of interest to this series include, but are not limited to, protein expression and processing; nanotechnology; molecular engineering and computational biology; environmental sciences; food biotechnology, genomics, proteomics, and metabolomics; large-scale manufacturing and commercialization of human therapeutics; biomaterials and biosensors; and regenerative medicine. We expect these publications to be of significant interest to the practitioners both in academia and industry. Authors and editors are carefully selected for their recognized expertise and their contributions to the various and far-reaching fields of biotechnology.

The upcoming volumes will attest to the importance and quality of books in this series. I thank the fellow coeditors and authors of these books for agreeing to participate in this endeavor. Finally, I thank Ms Anita Lekhwani, Senior Acquisitions Editor at John Wiley & Sons, Inc., for approaching me to develop such a series. Together, we are confident that these books will be useful additions to the literature that will not only serve the biotechnology community with sound scientific knowledge but will also inspire them as they further chart the course of this exciting field.

Anurag S. Rathore
Amgen, Inc.

Thousand Oaks, California
January 2009

CONTRIBUTORS

Milton J. Axley, MedImmune, Gaithersburg, Maryland

Amit Banerjee, Pfizer Corporation, Chesterfield, Missouri

Jeffrey Carter, GE Healthcare, Westborough, Massachusetts

Douglas J. Cecchini, Biogen Idec, Cambridge, Massachusetts

Jean Harms, Amgen Inc., Thousand Oaks, California

Carol F. Kirchhoff, Pfizer Corporation, Chesterfield, Missouri

Steven Kozlowski, Food and Drug Administration, Silver Spring, Maryland

Maureen Lanan, Biogen Idec, Cambridge, Massachusetts

Duncan Low, Amgen Inc., Thousand Oaks, California

Rohin Mhatre, Biogen Idec, Cambridge, Massachusetts

Michael Molony, Allergan Corporation, Irvine, California

David L. Narum, National Institutes of Health, Rockville, Maryland

Kingman Ng, Eli Lilly and Company, Indianapolis, Indiana

Cynthia N. Oliver, MedImmune, Gaithersburg, Maryland

Joseph Phillips, Amgen Inc., Thousand Oaks, California

Natarajan Rajagopalan, Eli Lilly and Company, Indianapolis, Indiana

Kripa Ram, MedImmune, Gaithersburg, Maryland

Anurag S. Rathore, Amgen Inc., Thousand Oaks, California

John Rozembersky, Rozembersky Group, Inc, Boxborough, Massachusetts

Mark A. Schenerman, MedImmune, Gaithersburg, Maryland

Satish K. Singh, Pfizer Inc., Chesterfield, Missouri

Gail Sofer, Consultant, SofeWare Associates, Austin, Texas

Patrick G. Swann, Food and Drug Administration, Silver Spring, Maryland

Cenk Undey, Amgen Inc., West Greenwich, Rhode Island

Pim van Hoek, Amgen Inc., Thousand Oaks, California

Xiangyang Wang, Amgen Inc., Thousand Oaks, California

Gail F. Wasserman, MedImmune, Gaithersburg, Maryland

Peter K. Watler, JM Hyde Consulting, Inc., San Francisco, California

Keith Webber, Food and Drug Administration, Silver Spring, Maryland

1

QUALITY BY DESIGN: AN OVERVIEW OF THE BASIC CONCEPTS

Rohin Mhatre and Anurag S. Rathore

1.1 INTRODUCTION

The premise of Quality by Design (QbD) is that the quality of the pharmaceutical product should be based upon the understanding of the biology or the mechanism of action (MOA) and the safety of the molecule [1]. The manufacturing process should then be developed to meet the desired quality attributes of the molecule, hence the concept of "design" of the product quality versus "testing" the product quality. Although testing the product quality after manufacturing is an essential element of quality control, testing should be conducted to confirm the predesired product attributes and not to simply reveal the outcome of a manufacturing process. The ICH Q8 guideline provides an overview of some of the aspects of QbD [2]. The guideline clearly states that *quality cannot be tested into products*; that is, *quality should be built in by design*.

Although the task of designing a complex biological molecule such as a monoclonal antibody may seem daunting, the experience gained in the past roughly 30 years of the biotechnology industry history has laid the foundation for the QbD initiative [3, 4]. The industry has come a long way in identifying and selecting viable drug candidates, in developing high-productivity cell culture processes, in designing purification processes that yield a high-purity product, and in analyzing the heterogeneity of complex

Quality by Design for Biopharmaceuticals, Edited by A. S. Rathore and R. Mhatre
Copyright © 2009 John Wiley & Sons, Inc.

biomolecules. As all these activities are the building blocks of QbD, the concept of QbD has in fact been practiced for the last few years and has in turn led to the development of highly efficacious biopharmaceuticals and robust manufacturing processes. The issuance of the ICH Q8 guideline was an attempt to formalize the QbD initiative and to allow manufacturing flexibility based on the manufacturer's intricate knowledge of the molecule and the manufacturing process. The concept of obtaining intricate knowledge of the molecule along with the manufacturing process and the resulting flexibility in manufacturing, the eventual goal of the QbD initiative, requires an understanding of the various elements of QbD.

The two key components of QbD are [4]

1. The understanding of the *critical quality attributes* (CQAs) of a molecule. These are the attributes of the molecule that could potentially affect its safety and efficacy profile.
2. The *design space* of the process defined as the range of process inputs that help ensure the output of desired product quality.

An overview of these components is discussed further in this chapter and elsewhere in this book.

1.2 CRITICAL QUALITY ATTRIBUTES

The starting point of QbD is developing a good understanding of the molecule itself. Biomolecules are quite heterogeneous due to the various post-translational modifications that can occur and have been commonly observed. These modifications arise from the glycosylation, oxidation, deamidation, cleavage of labile sites, aggregation, and phosphorylation, to name a few. As many of these modifications could impact the safety and efficacy of the molecule, defining the appropriate CQAs of the molecule is an important starting point in the development cycle of a biopharmaceutical. Although the understanding of the CQAs evolves during the life cycle of the product, understanding the CQAs at an early stage of the development of the molecule is clearly desirable. Studies conducted during the early research stages of development of a potential biopharmaceutical may entail evaluating various forms of a particular biomolecule in animal studies. The outcomes of such studies help "design" a biomolecule with the desired quality attributes so as to be safe and highly efficacious.

Since the CQAs can impact the safety and clinical efficacy of a molecule, data gathered in animal studies, toxicological studies, and early human clinical trails become the starting point for defining the CQAs. On the basis of the safety and efficacy readout of a clinical trial, one can start to define the product profile of a molecule. The assumption is that if the CQAs of the molecule are similar to those used in preclinical and clinical trials, the safety and efficacy will be comparable as well. Furthermore, historical data from clinical trials of similar molecules can also provide valuable insight into the CQAs. Evaluation of the *in vitro* biological activity via bioassays, reflecting the mechanism of action, can provide a good assessment of how the various product attributes could potentially impact the *in vivo* activity of a molecule. The molecule can be altered by conducting stress studies to induce higher level of aggregation; oxidation, deamidation, and the glycosylation pattern can be varied as well. The impact of changes in the molecular structure on the biological activity can then be evaluated via various bioassays. This study is referred to as structure–activity relationship. The evaluation of *in vitro* activity is often the relatively easiest means of determining the CQAs. However, *in vitro* assessments can only provide an understanding of the potential changes in the activity of the molecule, and the correlation between this change in activity and the impact of efficacy in patients is often unclear. Further assessment of the molecule in animal studies to evaluate clearance, efficacy, and safety is often a good indicator of the behavior of the molecule in human trials and is a better tool for understanding the CQAs. Additional details of the determination of the CQA can be found in Chapter 4.

1.3   AN OVERVIEW OF DESIGN SPACE

After defining the CQAs, the next and more critical step is the development of a manufacturing process that will yield a product with the desired CQAs [4, 5]. During the process development, several process parameters are routinely evaluated to assess how they could impact product quality [6]. The design space for the process eventually evolves from such a study. For example, during the cell culture development, ranges for process inputs such as temperature, pH, and the feed timing can be evaluated to determine if operating within a certain range of temperature and pH has an impact on product quality. The design of experiments (DOE) is conducted in a manner so as to evaluate the impact of the multiple variables (multivariate) and also to understand if and how changes in one or more of the process inputs have an effect on the product quality and/or if a process input is independent of changes in other inputs.

The design space (process range) is then established for each of the above process inputs. This can be further explained using an example of a design space for a purification column. If a column used to purify a protein is expected to reduce the level of protein aggregate to 2%, the various column operating parameters such as flow rate, pH and strength of the buffer, load volume, and so on are evaluated such that operating within a certain range of these parameters yields an aggregate level of less than or equal to 2%. If it turns out that pH above 6 or below 4 results in aggregate levels above 2%, then the design space for the pH of the buffer is defined as between 4 and 6. One can similarly envision a design space for the flow rate and other inputs for the particular purification step. Eventually, the entire production process for a molecule will have a defined design space, and operating within that design space should lead to a product of acceptable quality. Operating beyond the design space of a particular process input may result in an unacceptable product quality.

Since the production process for a biomolecule entails multiple steps starting from the cell culture process to the final purification and eventually to the formulation and fill in the desired container, the development of a design space for a particular step is not usually independent of other steps in the production process. Since the output of one step becomes the input for the next step, the development of the design space for a process should be evaluated in a holistic manner. One such approach may be to determine the desired product quality from the final process step and to work backward in the process to ensure that each step of the process delivers the required product quality needed for the next step to meet the quality target of the final step. To provide an example of this approach, we can revisit the above example of a desired level of 2% aggregate in the final drug product. In this particular case, design space should be developed for all parameters of the production process that can potentially impact the level of aggregate in the final drug product. The maximum level of aggregate resulting from each of the steps in the process does not need to be less than 2%, particularly steps that are upstream in the process such as the protein A purification, the first columns used in the purification of an antibody. The development of the design space including the design of experiments is discussed in detail in Chapters 5–7.

1.4   RAW MATERIALS AND THEIR IMPACT ON QbD

In addition to the design space and CQAs, other factors also play an important role in implementing QbD, and raw material is one such factor. Cell culture processes used to make recombinant proteins use complex growth media such as hydrosylates and also feeds such as vitamins for the cell. The understanding of how the various components of these complex raw materials affect the productivity of the cells and the quality of the product is not a trivial task. It requires a thorough analysis and quantitation of the various components of the raw material. Raw material analysis and correlation between raw material components and the productivity of cells and product quality is an area that has not been sufficiently explored by the biotechnology industry. However, the evolution of instruments such as high-resolution nuclear magnetic resonance spectroscopy, near infrared spectroscopy, and mass spectrometry has provided an opportunity to analyze complex mixtures of raw materials. In addition, the availability of sophisticated statistical tools for deconvolution and pattern matching of complex data sets has further refined the approach to analyzing raw materials. Once the correlation between the critical components of the growth or feed media and the performance of cells is understood, the ultimate goal of raw material analysis in the context of QbD would be to fortify the media as needed with the relevant component so as to ensure the desired productivity and product quality. Further details of analysis of complex raw materials are provided in Chapter 11.

1.5   PROCESS ANALYTICAL TECHNOLOGY

Since one of the goals of QbD is to maintain control of the process to achieve the desired product attributes, process analytical technology (PAT) is an important tool for QbD. PAT entails analysis of product quality attributes during the various stages of the manufacturing process of a biomolecule. The analysis is often conducted online using either probes inserted into the bioreactor to monitor critical components such as the cell density or sterile sampling devices to divert the stream from a purification column to assess the product purity [7]. In either case, the online analysis enables operators on the manufacturing floor to make real-time adjustments to the process parameters so as to obtain the desired product profile at every stage of the manufacturing process. For example, a PAT tool to monitor a purification column would entail periodically sampling the elution stream from the column via a sampling device and diverting the sample to an online HPLC system [8]. The results of the online HPLC analysis, indicative of the product purity, would be used to determine the eluate volume that should be collected. In this particular example, the fraction of the eluate of purity below a predetermined criterion would provide a trigger to stop collection of the eluate and to divert the elution stream to waste. The advantage of such a PAT tool would be that the collection of the column eluate would be based on the required product purity and would help to ensure a consistent product quality for every production batch of the biomolecule [8]. Further applications of PAT can be found in Chapters 12 and 13.

1.6   THE UTILITY OF DESIGN SPACE AND QbD

Prior to development of the design space, the questions to ask are the following: How would the design space be used? What is the advantage for a company of developing a design space for any of its products? What would be the driver for regulatory agencies to promote the concept of design space and QbD?

As seen in Fig. 1.1, limits that establish the acceptable variability in product quality and process performance attributes would also serve as the process validation acceptance criteria [4, 5]. After the design space has been established, the regulatory filing would include the acceptable ranges for all key and critical operating parameters (i.e., design space) in addition to a more restricted operating space typically described for pharmaceutical products. After approval, CQAs would be monitored to ensure that the process is performing within the defined acceptable variability that served as the basis for the filed design space. The primary benefit of an expanded design space would be a more flexible approach by regulatory agencies. Process changes are often driven by changes in the manufacturing equipment and raw materials, to name a few. At present, changes in the process require formal filings and approvals from regulatory agencies and often require a significant commitment of both time and resources for the industry and the regulatory agencies. The outcome of the design space development (as stated in the ICH Q8 guideline) would be that upon the approval of the design space for a particular product by a regulatory agency, process changes within the design space would not require additional regulatory filing and approval. This shift in the paradigm of using enhanced process knowledge to enable process changes with a limited burden of regulatory approval is clearly beneficial to both the manufacturer and the regulatory agencies. Chapter 2 further reviews the regulatory relief and implications of the QbD initiative.

Process improvements during the product life cycle with regard to process consistency and throughput could take place with reduced postapproval submissions. As manufacturing experience grows and opportunities for process improvement are identified, the operating space could be revised within the design space without the need

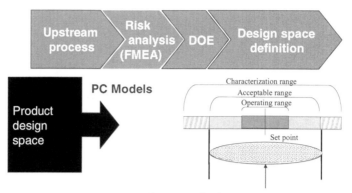

Process design space: key and critical
operational parameter values where all
resulting CQAs are within the pre-
established product design space

Figure 1.1. Illustration of the creation of process design space from process characterization studies and its relationship with the characterized and operating spaces. The operating range denotes the range in the manufacturing procedures and the characterization range is the range examined during process characterization. The acceptable range is the output of the characterization studies and defines the process design space. Adapted from Ref. [5], by permission of Advanstar Communications.

for postapproval submission. This is illustrated in Fig. 1.2, which shows that if the process creeps outside the design space, process changes may be required to be made and may require process characterization, validation, and filing of the changes to the approved process design space.

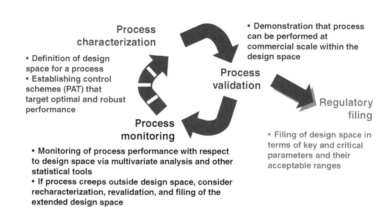

Figure 1.2. Application of the design space concept in process characterization, validation, monitoring, and regulatory filing. Adapted from Ref. [5], by permission of Advanstar Communications.

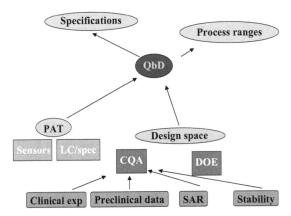

Figure 1.3. The various elements of QbD. The boxes in the bottom row show all the relevant information that is used to develop the critical quality attributes. The CQA and DOE data are then used to develop the design space. The design space and PAT tools help establish QbD.

1.7 CONCLUSIONS

Figure 1.3 depicts the various components of QbD discussed above and the correlation between the various components. As shown in the figure, the outcome of the QbD exercise is the establishment of the design space for the process and the operating ranges (ORs) that help achieve the desired product quality. As mentioned earlier, the reader is referred to the various sections of the book to gain further understanding of the various aspects of QbD. The editors hope that this book will help establish a good framework for any researcher to build Quality by Design into a manufacturing process for a biomolecule.

REFERENCES

[1] PAT Guidance for Industry: A Framework for Innovative Pharmaceutical Development, Manufacturing and Quality Assurance. U.S. Department of Health and Human Services, Food and Drug Administration (FDA), Center for Drug Evaluation and Research (CDER), Center for Veterinary Medicine (CVM), Office of Regulatory Affairs (ORA), September 2004.

[2] Guidance for Industry: Q8 Pharmaceutical Development. U.S. Department of Health and Human Service, Food and Drug Administration (FDA), May 2006.

[3] Kozlowski S, Swann P. Current and future issues in the manufacturing and development of monoclonal antibodies. *Adv Drug Deliv Rev* 2006;58:707–722.

[4] Rathore AS, Winkle H. Quality by Design for Pharmaceuticals: Regulatory Perspective and Approach. *Nature Biotechnology* 2009;27:26–34.

[5] Rathore AS, Branning R, Cecchini D. Design space for biotech products. *BioPharm Int* 2007;36–40.

[6] Harms J, Wang X, Kim T, Yang J, Rathore AS. Defining design space for biotech products: case study of *Pichia pastoris* fermentation. *Biotechnol Prog* 2008;24(3):655–662.

[7] Munson J, Stanfiled CF, Gujral B. A review of process analytical technology (PAT) in the U.S. pharmaceutical industry. *Curr Pharm Anal* 2006;2:405–414.

[8] Rathore AS, Yu M, Yeboah S, Sharma A. Case study and application of process analytical technology (PAT) towards bioprocessing: use of on-line high performance liquid chromatography (HPLC) for making real time pooling decisions for process chromatography. *Biotechnol Bioeng* 2008;100:306–316.

2

CONSIDERATIONS FOR BIOTECHNOLOGY PRODUCT QUALITY BY DESIGN

Steven Kozlowski and Patrick Swann

2.1 INTRODUCTION

In August 2002, the Food and Drug Administration (U.S. FDA) announced a significant new initiative, pharmaceutical Current Good Manufacturing Practices (CGMPs) for the twenty-first century [1]. This initiative is intended to enhance and modernize pharmaceutical manufacturing and product quality. Specific areas of focus include facilitating industry adoption of risk-based approaches, technological advances, and modern quality management techniques. As part of this initiative, the FDA will use state-of-the-art pharmaceutical science in developing review, compliance, and inspection policies and will coordinate these activities under a quality systems approach.

Concurrently with the CGMPs for the twenty-first century initiative, process analytical technology (PAT), a system to improve pharmaceutical manufacturing was being discussed at the advisory committee of the Office of Pharmaceutical Science at CDER and at the FDA Science Board [2].

Process Analytical Technology is a system for designing, analyzing, and controlling manufacturing through timely measurements (i.e., during processing) of critical quality and performance attributes of raw and in-process materials and processes with the goal of ensuring final product quality [3]. In 2004, the FDA published guidance on PAT [4] that

Quality by Design for Biopharmaceuticals, Edited by A. S. Rathore and R. Mhatre
Copyright © 2009 John Wiley & Sons, Inc.

described multivariate tools for design, data acquisition and analysis, process analyzers and controllers, continuous improvement, and knowledge management tools.

Systematic approaches to pharmaceutical manufacturing may be of benefit even if they do not use each of these specific tools. Although PAT may allow for greater flexibility, manufacturing may still be improved in the absence of real-time analysis of material attributes and without real-time linkage to process control. The term Quality by Design (QbD) [5, 6] is used to describe a more general approach to systematic pharmaceutical manufacturing. As described by Dr. Janet Woodcock, the desired state that drives all these manufacturing initiatives is a maximally efficient, agile, flexible pharmaceutical manufacturing sector that reliably produces high-quality drug products without extensive regulatory oversight [7].

Over the last few years, there has been significant progress in moving forward with these initiatives for small molecules, including a pilot program for QbD submissions [8]. However, biotechnology products are a growing part of the drug development pipeline [9]. It is important to consider how to approach the "desired state" for more complex products, such as biotechnology products. The principles of Quality by Design should be applicable to all pharmaceuticals including biotechnology products [10].

2.2 QUALITY BY DESIGN

Quality by Design is defined as a systematic approach to development that begins with predefined objectives and emphasizes product and process understanding and process control based on sound science and quality risk management [11]. Dr. Moheb Nasr has summarized QbD in a diagram [12] (Fig. 2.1). A systematic approach to pharmaceutical development should start with the desired clinical performance and then move to product design. The desired product attributes should then drive the process design, and the process design should drive the strategies to ensure process performance. This systematic approach may be iterative and thus the circular design as shown in Fig. 2.1. The inner circle interacts with many other specific measures of pharmaceutical manufacturing, such as specifications, critical process parameters, and so on. This QbD circle can be divided into two major semicircles, product knowledge and process understanding. A critical tool for enabling QbD manufacturing is a defined way of linking these two semicircles.

The International Conference on Harmonization of Technical Requirements for Registration of Pharmaceuticals for Human Use (ICH) has bridged this gap using the concept of a design space. A design space is the multidimensional combination and interaction of input variables (e.g., material attributes) and process parameters that have been demonstrated to provide assurance of quality [3]. This is the scientific definition of a design space. Design space also has a regulatory definition. Movement within a design space is not considered as a change that requires regulatory approval. However, change within a design space does need oversight by the sponsor's quality system. Design space is proposed by the applicant and is subject to regulatory assessment and approval. A design space could potentially link process performance to variables such as scale and equipment. The design space is thus a very flexible tool that links process characteristics and in-process material attributes to product quality. A recent definition of product

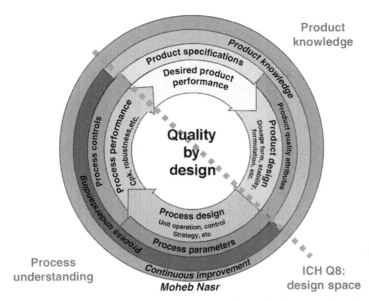

Figure 2.1. Quality by Design is a systematic approach to development that begins with predefined objectives and emphasizes product and process understanding and process control based on sound science and quality risk management (ICH Q8R1). Quality by Design begins by defining the desired product performance and also by designing a product that meets those performance requirements. The characteristics of the designed product are the basis for designing the manufacturing process, and the performance of the manufacturing process also needs to be monitored. Each of these steps may impact each other. For example, process performance may provide knowledge regarding manufacturability that could impact product design in an iterative manner. These steps also relate to specific quality measurements and tools. Product specifications should ideally be based on desired clinical performance. The product design should be defined in terms of product quality attributes. The criticality of these attributes and the relationship of these attributes to specifications may evolve over the product life cycle. Process parameters are important in defining the process, and process controls are important in ensuring process performance. This circle of QbD can be split into two general areas, product knowledge and process understanding. These two areas meet in the design space and the interaction of product knowledge and process understanding allows for continuous improvement. The QbD circle was developed by Dr. Moheb Nasr. (See the insert for color representation of this figure.)

quality was given by Dr. Janet Woodcock [13], "Good pharmaceutical quality represents an acceptably low risk of failing to achieve the desired clinical attributes." As indicated at the top of Fig. 2.1, QbD starts with the desired clinical performance.

2.3 RELEVANT PRODUCT ATTRIBUTES

In the draft annex to ICH Q8, ICH Q8(R1) [11], the target product profile is described as a starting point for Quality by Design. The target product profile is based on the desired clinical performance. A number of considerations for the target product profile are described including dosage form, strength, release characteristics, and drug product quality characteristics (e.g., sterility, purity). All of these are also considerations for biotechnology products; however, there may be a difference in focus for these complex products.

Currently, many protein products are delivered parenterally, although orally administered enzymes [14, 15] and novel dosage forms such as inhaled insulin [16] have been developed. Even for parenteral dosage forms, there are important considerations regarding formulation (e.g., liquid versus lyophilized) and route (e.g., intramuscular, subcutaneous, or intravenous). The choice of delivery system, such as prefilled syringes, is also important. For complex as well as for simple drugs, desired pharmacology, targeted patient population(s), and disease state(s) should be carefully considered in decisions on drug product dosage form, strength, and route of administration.

For biotechnology products, the complex processes, raw materials, and biological substrates used in manufacturing can lead to a broad range of process-related impurities. These impurities may impact product performance beyond direct toxicities. Impurities may impact the attributes of the active pharmaceutical ingredient, such as protein aggregation by tungsten moieties [17]. Contaminating proteases may also impact stability of a protein product. Some impurities can act as adjuvants and thus may have the potential to alter protein immunogenicity. Such impurities have been suggested as playing a role in erythropoietin immunogenicity [18, 19] although other possible causes have also been suggested [20, 21].

Product quality characteristics for complex products encompass a wide variety of product variants that include product-related substances and product-related impurities [22]. The three-dimensional structure of proteins is important for receptor interactions. Changes in folding may alter receptor binding and/or signaling. Protein multimerization may change receptor blockade to receptor activation. Abnormally folded proteins may impact immunogenicity through aggregation, generation of novel epitopes, and/or altered uptake by antigen presenting cells. Protein folding depends on low-energy interactions such as hydrogen bonds. Thus, minor environmental changes could impact protein structure and generate structural variants. Environmental excursions during processing or shelf life may interact with other impurities such as trace proteases and impact degradation.

In addition to variants in higher order structure, proteins can have many post-translational modifications. Biotechnology products are often heterogeneous mixtures with many variants that have different sets of modifications. Although many post-translational modifications may not impact product performance, others may alter pharmacokinetics, activity, or safety (e.g., immunogenicity). In Fig. 2.2a, a schematic of a monoclonal antibody is shown with a subset of potential post-translational modifications, such as N-terminal pyroglutamines, oxidations, deamidations, glycations, C-terminal lysines, and glycosylations. One of these modifications, an oxidation site, is considered in a decision tree regarding a potential impact on performance (Fig. 2.2b). Such a decision tree can be assigned for each of the modifications. Decision trees can also be constructed regarding specific safety concerns, such as immunogenicity. Ideally, a probability or risk ranking can be applied to the various possibilities in each of

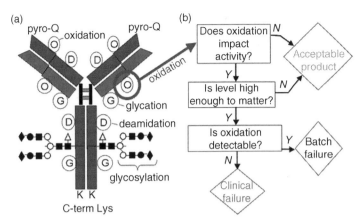

Figure 2.2. Biotechnology products have a large number of structural attributes that may impact product performance. (a) Some structural attributes of monoclonal antibodies are indicated, such as pyroglutamine, oxidation, glycation, deamidation, glycosylation, and clipping of C-terminal lysines. An oxidation at one site is circled in blue. (b) A decision tree for the impact of that oxidation on product activity is shown. (See the insert for color representation of this figure.)

these decision trees. For example, immunogenicity is difficult to predict on the basis of attributes, but the impact of immunogenicity can be evaluated in terms of clinical risk [23–25]. Risk assessments, whether for activity or safety, are challenging and require the use of many sources of information such as prior knowledge, related product or platform data, *in vitro* and *in vivo* biological characterization of product variants, and clinical data. In many cases, no one source of information would allow a meaningful risk assessment. However, the integration of multiple sources of information may facilitate a useful assessment of risk. A matrix approach can be informative [26], evaluating product lots generated at varying points throughout the development for biological impact across a variety of studies. This information can give confidence to proposed mechanisms of action and structure–function relationships. These data may be integrated around mechanism of action models and potentially use Bayesian statistical approaches.

For products that share significant sequence homology to related products, a platform approach to attribute impact may be very useful. Monoclonal antibodies may present opportunities [26] for this, but the nature of the targets, patient population, disease state(s), and the role of effector functions need to be considered in any extrapolations. Studies demonstrating that a product attribute is rapidly modified *in vivo* (e.g., deamidation or oxidation of a specific residue) may allow for increasing levels of that attribute over product shelf life. Comparisons to related endogenous molecules may also be informative.

Clinical studies on product variants would provide the strongest linkage between product attributes and clinical performance. In this book [27], an example of using structural characterization of variants in timed samples for variant pharmacokinetics is given. Such studies do not require purification of variants and additional clinical studies. These studies may allow the evaluation of variant effects on pharmacokinetics and, in some cases, variant effects on pharmacodynamics. In addition, during standard clinical development many different lots may be used at different times after manufacture. The use of multivariate statistical analysis for evaluating product attributes and clinical findings may be informative [28].

There are many potential strategies for understanding the input of complex product attributes on clinical performance. Attributes may interact and a multivariate attribute space may be more useful than univariate ranges for such attributes. However, complex products may have many thousands of possible attribute states. Even using all these strategies, this large number of attributes and interactions cannot be fully evaluated. Thus, a risk assessment, utilizing prior knowledge in conjunction with other approaches, will be needed. An important consideration for complex products is that the more the attributes dismissed based on assumptions, the greater the uncertainty that no critical attribute or interaction was missed. This greater uncertainty needs to be considered in defining attributes as unimportant. For complex products, mechanism of action and biological characterization may contribute to understanding the importance of product attributes.

We have discussed assigning risk and importance to many attributes of complex products. For regulatory purposes, clear ways to distinguish attributes will be needed. Draft guidance [11] has defined a critical quality attribute (CQA) as a physical, chemical, biological, or microbiological property or characteristic that should be within an appropriate limit, range, or distribution to ensure the desired product quality. This CQA definition is process independent and does not consider the control system; a CQA is only defined as a property or characteristic of the product. Other suggestions have included an intermediate level of attribute importance, analogous to intermediate levels of parameter importance or key parameters [29]. For complex products with greater uncertainties regarding attribute importance, intermediate attribute categories, between noncritical and critical, may be of value. Linkage of potential attribute risk to the appropriate control strategy may be best achieved with a variety of attribute risk categories. Attribute importance can be assigned through a risk assessment, as described above. CQAs do not necessarily need to be controlled by end-product testing (classical specifications), but the sensitivity of biotechnology product attributes to environmental conditions should be considered in any upstream approaches to CQA control.

Assessment of product attributes using the above methods may require significant efforts. However, the knowledge gained from linking product attributes to clinical performance can be leveraged into other products and may facilitate discovery and design of new products. This information does not need to be complete at the time of marketing application submission. As product knowledge grows, CQAs may evolve. Not all sponsors may invest in an extensive risk assessment of all their products before approval. Historically, sponsors have used the ranges of attributes in clinical study lots to assure clinical performance. This approach may still be appropriate in some cases but could limit flexibility.

2.4 MANUFACTURING PROCESS

In Fig. 2.3, a schematic for a typical biotechnology manufacturing process is described. The upstream process begins with expansion of the cell substrate that produces the

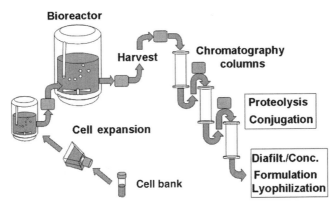

Figure 2.3. A schematic for manufacturing of a biotechnology product. The manufacture of a biotechnology product generally involves many steps. Manufacture often being with the thawing of frozen cells then continues through a series of cell culture expansions into the final production bioreactor. A harvesting step to remove cells and other culture components precedes downstream purification. For purification, the product then undergoes a series of purification steps, such as affinity, ion-exchange, hydrophobic, or size exclusion chromatography. The purified bulk product is then concentrated and formulated and may be lyophilized. Additional processing steps such as proteolysis or conjugation may occur as part of manufacturing. These steps would generally be followed by some purification to remove step residuals.

product and continues through the final production cell culture to the harvesting of unpurified bulk product. Downstream process steps include initial purification (e.g., chromatography), modifications (e.g., conjugation), and final polishing steps leading to the bulk drug substance. Filling and/or lyophilization of the drug substance are then performed. Many of the important attributes of a biotechnology product are impacted by the upstream manufacturing. The cells used to manufacture the product are miniature factories affected by many variables including subtle differences in media composition, aeration, metabolites, shear forces [30], and cell density. Differences in these factors can impact a wide variety of product structural attributes including glycosylation, oxidation, cellular proteolytic processing, and so on. Many challenging process impurities are generated during upstream manufacturing, such as host cell proteins, DNA, and media components. Systems biology approaches to clonal selection or alteration [31], cellular metabolism, and biosynthesis along with better defined media components, improved models for bioreactor fluid dynamics [32], more sophisticated monitoring (e.g., beyond pH and dissolved O_2), and multivariate statistical analysis [33] may improve the understanding associated with these complex process steps. An improved process understanding linking process parameters to important product attributes could allow for a better design of an upstream process. Improved process understanding could also allow for broad design spaces and opportunities for changes in scale, equipment, and so on without prior FDA approval.

Although there are many potential opportunities for upstream processing, many challenges remain in understanding such complex process steps. Downstream processing for biotechnology products may afford more immediate opportunities for generation of large design spaces. To illustrate approaches to a downstream unit operation, a hypothetical column purification step will be considered. The use of a design space approach is compared to more traditional process limits in Fig. 2.4. Although manufacturing has always had ranges for operating parameters, they have generally been univariate. The pH and protein load of a chromatography column are used as example parameters. The linkage of these parameters to quality has been empirical, often based on the limited ranges used during manufacture of clinical trial material (Fig. 2.4a). Complex biological products have been defined by their manufacturing processes [34]; the process is the product. Changes in the manufacturing process often required a clinical trial to maintain the empirical link between process characteristics and product quality. By the mid-1990s [35], a specified subset of well-characterized biological products could utilize biochemical comparability to allow some manufacturing changes in the absence of new safety and efficacy studies. However, the parameter ranges used in manufacturing clinical or to-be-marketed material are often limited and do not explore the full extent of the ranges leading to acceptable product quality. Small-scale studies can be used to support wider parameter ranges (Fig. 2.4b). However, the validity of the scale-down models needs to be demonstrated and the column performance measurements used should link to product performance. Although small-scale studies can expand univariate ranges, they do not

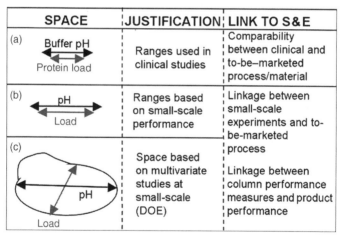

Figure 2.4. Spaces or ranges that can be used in product manufacturing. (a) Biologic and biotechnology products were historically defined by their manufacturing process. The process and process ranges were based on the ranges used for the product used in clinical trials showing safety and efficacy (S&E). Comparability studies allowed for some process changes. (b) These ranges were often narrow and could be expanded in small-scale models. The use of these wider ranges in manufacturing was dependent on the validity of the scale-down models and the performance criteria. (c) Since many variables interact, a multivariate space is more reflective of reality and may be generated using design of experiments. Generally, these experiments are also done at small scale and depend on the validity of the DOE models as well as the scale-down model and performance criteria.

account for interactions between parameters. Multivariate experiments at small scale can define a more meaningful space for parameters that may interact (Fig. 2.4c). Such a space can be generated with an efficient number of experiments using design of experiment (DOE) approaches [36, 37]. As with univariate small-scale experiments, the validity of the scale-down and the linkage of measured responses to product performance should be described. In addition to these considerations, the modeling used in DOE approaches should be justified.

An important advantage of generating a multidimensional design space is described in Fig. 2.5. An empirically derived manufacturing process is static with locked-in parameters (Fig. 2.5a). Any variability in process inputs is transferred to the product

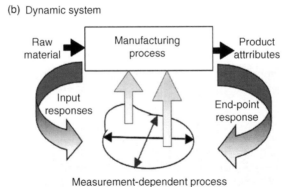

(a) Traditional paradigm

Variability

Raw material → Manufacturing process → Product attrributes

Locked process variables

(b) Dynamic system

Raw material → Manufacturing process → Product attrributes

Input responses

End-point response

Measurement-dependent process

Figure 2.5. A design space allows for a dynamic approach to manufacturing that transfers variability from important product attributes to process parameters. (a) The traditional paradigm for pharmaceutical manufacturing utilizes a fixed process. Thus, any variability in inputs may result in variable product. (b) In a dynamic manufacturing process, input variability can be monitored either directly or through product impact. Information regarding this variability can then be used to adjust the process parameters to compensate and produce high-quality product. A knowledge-rich design space that has been explored during development and throughout the product life cycle will allow the process flexibility necessary to compensate for variable process inputs. This figure was adapted from a diagram by Jon Clark.

since there is no way to compensate. In a dynamic process, information on variability of inputs or outputs can be used to tune the process (Fig. 2.5b). This information can be real time or based on off-line testing. Variable input parameters can then be compensated for. In addition to process monitoring, a dynamic manufacturing system needs flexibility in setting process parameters or a design space. To allow for process adjustments, the design space also needs to predict how movement within the space will impact product attributes. A simple example of compensating for variability, using the design space shown in Fig. 2.4c, is as follows. The product yield from an upstream process was higher than expected, and material would need to be discarded based on protein load limits. However, adjusting the pH could support a higher protein load in the upper left corner of the design space. This would facilitate taking advantage of variability in the upstream process.

A design space can be generated for one unit operation, as done for the chromatography example above, or for an entire process [11]. In an entire process, such as shown in Fig. 2.3, final product quality attributes can drive the design of the final manufacturing step and process inputs for the final step can drive the outputs of the preceding step. This approach to unit operations can continue upstream until it defines acceptable inputs and outputs from the initial thawing of a cell bank vial.

2.5   DEVELOPING A DESIGN SPACE

In the previous section, a hypothetical design space for chromatography was presented. Figure 2.6 describes some of the initial steps for developing such a design space. It is important to first define the outputs that the manufacturing step will need to achieve. These will be the responses evaluated in studying the chromatography step. This requires establishing the CQAs that the final product must meet. After that, the chromatography output performance measures must be set so that the complete process will deliver the final product CQAs. The general requirements of the manufacturing step should also have been considered in the initial process design (e.g., choice of the methodology).

In Fig. 2.6a, the step requirements are removal of subpotent charge variants and impurities. A flow-through ion-exchange column is then chosen to achieve these requirements (Fig. 2.6b). The potential factors that could impact ion-exchange performance are listed in a cause and effect diagram, such as the fishbone or Ishikawa diagram (Fig. 2.6c). The list of factors should be extensive as not to miss any variables. Since not all the factors can be studied, an assessment of relative risk is performed. A failure modes and effects analysis (FMEA) is one such risk assessment and management tool (Fig. 2.6d) that considers the probability, severity, and detectability of an event [38, 39]. Appropriate cross-discipline expertise [40] and other prior knowledge are important in generating a meaningful risk assessment.

After the risk assessment, a more limited set of variables can be studied using DOE. In addition to defining the variables to be studied, appropriate ranges need to be set. For screening, the ranges should be set above experimental noise and wide enough to detect variables that matter. Screening approaches, such as fractional factorial or

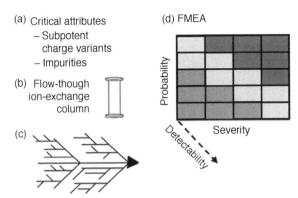

Figure 2.6. Generation of a design space starts with defining appropriate responses, the process methodology, and important process variables. (a) Charge variants and impurities are defined as important product attributes based on a risk assessment, such as the decision tree described in Fig. 2.2. (b) Based on prior knowledge, a flow-through ion-exchange purification step is felt to be an appropriate strategy for controlling these attributes in the product. (c) A fishbone or Ishikawa diagram can be used to describe a broad variety of variables and the relationship of the variables to product quality. (d) A risk assessment such as FMEA can be used to rank the importance of these factors based on probability, severity, and detectability. High risk is indicated in red, moderate risk in yellow, and low risk in green. (See the insert for color representation of this figure.)

Plackett–Burman designs, may be used to reduce the number of variables to be studied in detail [36, 37, 41]). These screening designs may confound different variable interactions or even confound main effects with interactions. After the further reduction of important variables by screening, a more extensive DOE may be used to refine estimates of variable effects and interactions. In Fig. 2.7, one such design, a central composite circumferential design (CCC), is shown. This design assumes that the risk assessment and screening studies have defined pH and protein load as the important factors in the flow-through ion-exchange step. In addition to assessing all the interactions at high and low levels (full factorial), the design includes multiple center points to assess experimental variability and extended axial points for each parameter (the other parameter is held at a midpoint value). The interpretation of DOE depends on modeling the relationship between variables and responses. A more extensive design, such as CCC, may be used to generate contour plots linking variables and responses. Such plots can be the basis of a design space. In Fig. 2.8a, three hypothetical response contour plots for basic variants, acidic variants, and an impurity are shown. On the basis of the need to achieve product CQAs, the acceptable limits for all three responses are indicated above the plots. Overlaying the acceptable areas for all three responses can result in a two-factor design space (Fig. 2.8b).

The use of DOE for chromatography of biotechnology products is not a new approach [42]. DOE has been used to demonstrate robustness within the empirical chromatography parameters. DOE can also be used to optimize chromatography parameters. For QbD, DOE may be used to establish an initial design space that extends beyond the empirical ranges used for pilot and full-scale lots. To support a

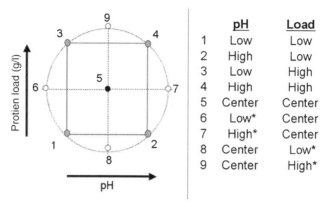

	pH	Load
1	Low	Low
2	High	Low
3	Low	High
4	High	High
5	Center	Center
6	Low*	Center
7	High*	Center
8	Center	Low*
9	Center	High*

Figure 2.7. A design of experiments for two variables. This is a central composite circumferential design, a potential design for generation of a response surface plot. The numbers associated with experimental points on the left are listed on the right in a table format indicating the experimental conditions. The axial points (6–9) extend beyond (1.4-fold) the interaction points (1–4), and this is indicated by a star in the table on the right.

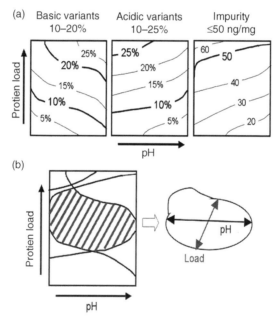

Figure 2.8. Generation of an initial design space from three response contour plots. (a) The anion exchange column step is intended to control charge variants and impurities. Hypothetical contour plots modeled from DOE studies for three important responses are shown. The limits of these responses based on clinical experience and appropriate studies are indicated above each plot. (b) An overlay of the acceptable areas of the three response plots can be used to define an initial design space. As indicated above, the data used to construct the design space are hypothetical and may not reflect the behavior of any product in real chromatography.

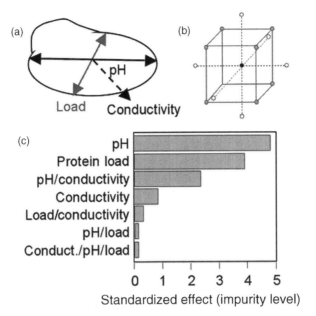

Figure 2.9. DOE often involves many variables, and the importance of each variable and interaction should be ranked. (a) The design space for a flow-through chromatography step may also need to consider conductivity. (b) A central composite circumscribed design for three variables is shown. (c) A Pareto chart can be used to assess the relative importance of main effects and interactions. For each of these, the response (e.g., impurity level) can be normalized and assigned significance on the basis of the variability of replicates and modeling residuals (deviations from model). The effects described are hypothetical and may not reflect the behavior of any product in real chromatography.

broad design space, wider ranges for variables can be set. However, ranges so wide that multiple experiments fail may require additional experiments with narrower ranges.

It is likely that more than two variables will be needed in developing a design space. The addition of conductivity to the design space is shown in Fig. 2.9a. A CCC design for three variables is shown in Fig. 2.9b. In Fig. 2.9c, a Pareto plot is used to rank the impact of each variable and all the possible interactions. Analysis of variance (ANOVA) can be used to assign significance to the effect of variables and interactions. In this hypothetical scenario, conductivity at the range explored has borderline significance. However, the interaction of pH and conductivity has a significant impact on the impurity level.

Considerations for acceptable DOE [36] may include replicate variability versus experimental ranges, appropriate transformations of data and/or results, appropriate choice of models (e.g., main effects, interaction, quadratic), choice of regression (e.g., partial least squares, multiple linear regression), goodness of fit (R^2, adjusted R^2, Q^2), and ANOVA (regression fit, lack-of-fit test).

2.6 UNCERTAINTY AND COMPLEXITY

A design space has regulatory implications. Movement within a design space is not considered a change [3]. Such movement can occur without a regulatory submission. However, movement within a design space does not remove responsibility from the sponsor's quality system [43, 44] to ensure product quality.

There can be many different spaces in a manufacturing unit operation (Fig. 2.10). The knowledge space is the space explored by small-scale DOE. The design space is the space explored by DOE that leads to acceptable product. A section of the design space may have been used to successfully manufacture pilot and full-scale lots. A smaller operating space will generally be used for initial manufacturing based on pilot and full-scale experience.

There are many possible sources of uncertainty in a design space generated by DOE. DOE is based on a model, and predictions of results within the design space are only as good as the model. The experiments are done at laboratory scale, and extrapolation to full-scale manufacturing should be justified. The generation of a design space can miss important factors at a number of levels. Variables can be overlooked in the initial fishbone diagram. Screening studies may discard variables that have important interactions if they do not have a significant main effect. An example of this would be the chromatography variables, cycle number, and regeneration buffer pH. Over the tested ranges, each of these variables may not impact product quality and could be disregarded based on screening experiments. However, together they may have significant impact.

If only a small number of variables are removed based on prior knowledge and screening experiments, the uncertainty is low. In addition, some process steps may have

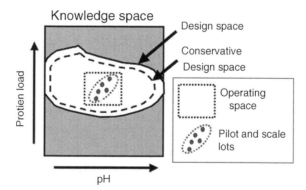

Figure 2.10. In developing a design space, there are many spaces to consider. The knowledge space represents the space explored by experimentation and may include areas of failures, indicated above in gray. The area labeled as design space includes the area modeled to produce acceptable product in small-scale experiments. A smaller operating space may limit experience at pilot or full-scale manufacturing. Uncertainty in risk assessments, scaling, modeling, and so on may limit confidence in the DOE defined space, and a more conservative design space may be appropriate in such situations.

mechanistic or first-principle models. These models could provide additional confidence regarding unimportant variables and facilitate scaling. For tablets, there is a long history of mechanistic or semimechanistic models for manufacturing unit operations such as roller compaction [45] and for tablet characteristics [46] that may impact performance. There are models for chromatography steps, but not all aspects may be covered (e.g., channeling through cracks, column behavior at the interface with the housing). Chromatography also depends on a variety of product-specific attributes [47]. Despite potential limitations, the presence of a relevant mechanistic model can increase confidence in a design space.

Pharmaceutical products can be placed along a continuum of complexity. Similarly, pharmaceutical manufacturing processes can also have differing levels of complexity. Biotechnology products tend to have both product attribute complexity and process complexity. Currently, there are only limited mechanistic models for either product attributes or complex process steps. Although biological characterization and knowledge of product mechanism(s) of action can improve the understanding of product attributes, there will still be uncertainty. Complex manufacturing steps may have many quantitative and qualitative factors and the potential for hidden interactions. Thus, creating design spaces for biotechnology products requires risk-based approaches to eliminate large numbers of variables. Even with advances in risk management, there is likely to be greater uncertainty for biotechnology product design spaces. There are approaches to deal with this uncertainty. In Fig. 2.10, a conservative design space is proposed. It may be prudent not to extend the design space to limits of failure when there is uncertainty regarding the design space. In addition, the quality system can be tailored to fit the complexity of the process [43, 44] and provide for appropriate risk management within design spaces. Although movement within a design space always needs be justified by the quality system, a design space with greater uncertainty may benefit from inclusion of a well-defined protocol to verify that movement within the design space does not have unintended effects on product quality. Protocols could focus directly on a risk-based verification that deals with areas of uncertainty. Inspections by regulators could check for the appropriate use of such protocols in managing significant movement within the design space. Minor changes to operating parameters and the ongoing verification of the initial design space should also be part of the sponsor's quality system. Verification at scale can be an important part of initial and ongoing process validation. The use of a multivariate approach to monitoring is likely to provide useful information over the product life cycle.

2.7 FUTURE HORIZONS

In the discussion of design space, we have used parameters such as protein load, pH, and conductivity. There are many more parameters that can be altered in a chromatography process step such as linear flow rate, buffer ions, washes, equilibration, and cycle number. All these parameters can be considered in developing the design space. For simplicity, all these are squeezed into the two-dimensional space in the lower left corner of Fig. 2.11. There are still many more dimensions for changes to this process step.

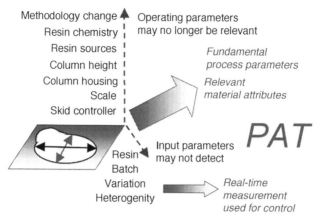

Figure 2.11. Moving beyond process parameters. The design space described in Fig. 2.10 and earlier is defined entirely by operating parameters that would be specific to the chromatography methodology and resin used. A wider range of potential changes is shown on the vertical axis extending from a design space. Dealing with such changes requires parameters that are less dependent on equipment, such as fundamental process parameters and relevant material attributes. In addition, even an excellent design space, based on appropriate risk assessment, experiments, and modeling, may not capture heterogeneity in inputs. Real-time measurements may capture input heterogeneity. However, real-time measurements of relevant material attributes are important characteristics of PAT.

The vertical axis lists some of these changes. These can vary from the chromatography skid and associated pumps and controllers, with easily defined performance criteria, such as pressure and flows, to a completely different methodology for removing impurities and inactive charge variants. In between these extremes are changes from column scale to resin chemistry. In addition to intentional process changes, there may be unexpected input changes that would be difficult to detect, such as heterogeneity in a batch of resin. For large changes in methodology, a design space of operating parameters would no longer be relevant. However, a design space of fundamental parameters, dimensionless variables, and ideally, relevant in-process material attributes may allow for such changes. For unexpected variability in inputs, real-time measurements could facilitate control. If in-process material attributes and real-time measurement control are added together, they equal PAT, a system for designing, analyzing, and controlling manufacturing through timely measurements (i.e., during processing) of critical quality and performance attributes of raw and in-process materials and processes [3]. PAT approaches would allow the maximum manufacturing flexibility.

However, for biotechnology products, process steps often have many functions. In our hypothetical chromatography example, three outputs were used to characterize the column performance, basic variants, acidic variants, and an impurity. In reality, a column could remove additional impurities and play a role in process viral clearance. All the important process step functions need to be considered in generating a design space. One means to identify the functions of a process step is to characterize the attributes of a product manufactured without that process step. For example, the absence of an ion-exchange chromatography polishing step may result in increased host cell proteins and aggregates and a shift in product charge variants.

Understanding the product attributes affected by a process step would also be important for a PAT approach. Therefore, multiple material attributes would usually need to be monitored. It may be more likely to have a design space that combines operating parameters with material attributes since not all important step functions may be evaluated by sensors. For very complex process steps such as cell culture, sensors for relevant material attributes or their surrogates could be very useful. The complexity and uncertainty regarding attributes of complex products suggest that even with monitoring of material attributes, movement within the design space will require risk-based verification by the sponsor's quality system. As described for the more limited chromatography design space, this verification should be specific to the uncertainties associated with a particular process step. Thus, there would need to be a risk- and knowledge-based approach to significant movement within a PAT-based design space. Also as described above, even in the absence of major adjustments to the process, verification of process performance is a life cycle issue. For PAT processes, the data from monitoring through a variety of sensors and parameters can be integrated using multivariate statistical analysis and/or control. In addition to verifying the design space, these data would be very useful in process improvement.

2.8   QbD SUBMISSION THOUGHTS

There has been discussion as to the organization of QbD information in submissions to regulatory agencies. There is no definitive approach for this; however, some possibilities can be considered. For small molecules, much of the QbD discussion has focused on the drug product. For biotechnology products, many complex process steps that can impact a large number of product attributes occur in manufacture of drug substance. Much of QbD information can fit into specific sections of the Common Technical Document (CTD) for registration of pharmaceuticals [48].

QbD information is best conveyed as a story from target profile, to CQAs, to process development and controls for each unit operation. A sequential (backward or forward) description of each unit operation would be helpful. Such a narrative should be located in one place. Possibilities include the manufacturing description or development sections, or an overarching narrative in Module 2: Quality Overall Summary. Ideally, the narrative would be sufficient to give review staff a clear picture of the product and process knowledge [12] and cite or hyperlink to the relevant detailed sections in other parts of the submission.

The content to support a QbD submission is also being discussed. The goal is to present knowledge, not raw data. However, there must be sufficient data to allow for understanding. Graphical representations should be accompanied by appropriate statistics and summary worksheets. Information based on models should describe the models used and evidence for their validity. Sources for prior knowledge and approaches to risk assessment and management should be well described. Initially, large amounts of data may be needed to support QbD applications for biotechnology products. As experience with QbD increases, the data in submissions may decrease; however, the relevant data should be available on inspections.

2.9 IMPLEMENTATION PLANS

It is clear that both agency and industry need experience to move forward with QbD for biotechnology products. The small-molecule world has made notable progress in implementing QbD through a variety of conferences, Cooperative Research and Development Agreements (CRADAs), and workshops [49] using mock products [50]. However, one of the most beneficial programs for moving forward has been the QbD pilot program by the Office of New Drug Quality Assessment (ONDQA) at CDER [8]. By encouraging industry to provide real examples, important issues were clarified. The use of risk assessments and DOE were important topics in QbD pilot submissions [51].

Although information from the ONDQA QbD pilot and other small-molecule experience is very relevant for biotechnology products, there are special considerations for complex products and processes. To deal with these considerations, a pilot for biotechnology products has been suggested and discussed at a number of venues [52, 53]. This pilot should focus on unique issues for biotechnology products and also provide information on applying QbD to both unit operations and entire applications. Since there are fewer biotechnology product applications than small-molecule applications, it would be useful to include supplements for already approved products in the pilot. Full QbD applications should ideally involve multiple unit operations that are linked, and such applications could benefit from earlier sponsor–agency interactions. In addition to the pilot, workshops discussing mock biotechnology QbD applications would also be of value.

One important issue that was highlighted by the small-molecule pilot was the need for an approach to regulatory opportunities. Industry is very interested in regulatory approaches to managing postapproval changes under QbD. One approach to this would take advantage of current regulations for protocols [54, 55]. Comparability protocols are a type of protocol used for manufacturing changes. Comparability protocols provide upfront information and criteria for a type of change and if approved, allow for a reduced reporting category for that change. Although comparability protocols have been used for multiple changes, they have often been used for a single change. By providing the enhanced product and process knowledge associated with QbD, a protocol could cover a broader range of changes. Such a protocol could be called an expanded change protocol and would be a useful tool in a biotechnology QbD pilot. The use of specific protocols by sponsor's quality systems in dealing with design space uncertainty was suggested earlier. Expanded change protocols could provide information on developing quality systems protocols for industry managed changes. The pilot could also provide information on approaches to site changes and inspectional issues for QbD submissions. For QbD submissions using an expanded change protocol, the protocol may be another possible location for the QbD narrative.

2.10 SUMMARY

There has been significant progress in developing Quality by Design approaches to small molecules. Many of the principles used for small molecules, such as design space, risk assessments, design of experiments, and PAT, apply to biotechnology products. However, biotechnology manufacturing involves both complex products and complex processes. This combined complexity leads to greater uncertainty for biotechnology product design spaces. Approaches to this uncertainty include a greater use of biological characterization, mechanism(s) of action, and structure/function analysis in risk assessments for product attributes. Appropriate statistical support for modeling, platform approaches, conservative boundaries, and risk-based strategies for movement within a design space can also provide confidence. Experience with actual applications in a pilot program could provide information on organization, content, and appropriate evaluation of QbD for biotechnology products.

A QbD application is not a requirement, but some level of product and process knowledge is generally necessary for complex products. Although many safe and effective products have been brought to market without demonstrating all the features of QbD, the proactive and systematic use of risk management and science in manufacturing will ultimately make safer, better products at lower cost. In addition, leveraging some of the structure/function information associated with QbD may have added benefits, such as facilitating clinical development and drug discovery.

Over the past two and a half decades, biotechnology-based pharmaceuticals have moved from a single marketed product to a prominent role in treatment of many diseases. Over this period, there have been many advances in the development, manufacturing, and regulation of these products [56]. QbD can be an important step forward in this evolution.

ACKNOWLEDGMENTS

The authors acknowledge Keith Webber for his valuable comments in review of this chapter. This chapter reflects the current thinking and experience of the authors. However, this is not a policy document and should not be used in lieu of regulations, published FDA guidance, or direct discussions with the agency.

REFERENCES

[1] U.S. FDA. Pharmaceutical cGMPs for the 21st Century: A Risk-Based Approach: Final Report, 2004.

[2] U.S. FDA. Science Board Executive Summary November 16, 2001.

[3] ICH. Q8: Pharmaceutical Development, 2005. http://www.fda.gov/cder/guidance/6746fnl.htm.

[4] U.S. FDA. Guidance for Industry PAT—A Framework for Innovative Pharmaceutical Development, Manufacturing, and Quality Assurance, 2004. http://www.fda.gov/cder/guidance/6419fnl.pdf.

[5] Scherzer R. Quality by Design: A Challenge to the Pharma Industry. Board FS, 2002.

[6] Hussain AS. Quality by Design: Next Steps to Realize Opportunities. Presented to the Food and Drugs Administration Advisory Committee for Pharmaceutical Science: Manufacturing Science Subcommittee, 2003.

[7] Woodcock J. Pharmaceutical quality in the 21st century—an integrated systems approach. Workshop on Pharmaceutical Quality Assessment—A Science and Risk-Based CMC Approach in the 21st Century, North Bethesda, MD, AAPS, 2005.

[8] U.S. FDA. Submission of chemistry, manufacturing, and controls information in a new drug application under the new pharmaceutical quality assessment system: notice of pilot program. *HHS, Federal Register* 2005;70:40719–40720.

[9] BIO. Biotechnology Industry Facts, 2008. http://bio.org/speeches/pubs/er/statistics.asp.

[10] Rathore AS, Branning R, Cecchini D. Quality: design space for biotech products. *BioPharm Int* 2007;20(4):36–40.

[11] ICH. Q8(R1) Pharmaceutical Development Revision 1, 2008. (step 4, 11/13/2008, http://www.ich.org/cache/compo/363-272-1.html#Q8).

[12] Nasr M. Risk-based CMC review and quality assessment: What is quality by design (QbD)? 2006 FDA/Industry Conference, School of Pharmacy—Temple University, 2006.

[13] Woodcock J. The concept of pharmaceutical quality. *Am Pharm Rev* 2004;7:1–3.

[14] Morrow JD. Pancreatic enzyme replacement therapy. *Am J Med Sci* 1989;298:357–359.

[15] Littlewood JM, Wolfe SP, Conway SP. Diagnosis and treatment of intestinal malabsorption in cystic fibrosis. *Pediatr Pulmonol* 2006;41:35–49.

[16] Jani R, Triplitt C, Reasner C, Defronzo RA. First approved inhaled insulin therapy for diabetes mellitus. *Expert Opin Drug Deliv* 2007;4:63–76.

[17] Vaczek D. Promoting dosing accuracy with pre-filled syringes. *Pharm Med Pack News* 2007; s42–s50.

[18] Boven K, Knight J, Bader F, Rossert J, Eckardt KU, Casadevall N. Epoetin-associated pure red cell aplasia in patients with chronic kidney disease: solving the mystery. *Nephrol Dial Transplant* 2005;20(Suppl. 3):iii33–iii40.

[19] Ryan MH, Heavner GA, Brigham-Burke M, McMahon F, Shanahan MF, Gunturi SR, Sharma B, Farrell FX. An *in vivo* model to assess factors that may stimulate the generation of an immune reaction to erythropoietin. *Int Immunopharmacol* 2006;6:647–655.

[20] Schellekens H, Jiskoot W. Eprex-associated pure red cell aplasia and leachates. *Nat Biotechnol* 2006;24:613–614.

[21] Locatelli F, Del Vecchio L, Pozzoni P. Pure red-cell aplasia "epidemic"—mystery completely revealed? *Perit Dial Int* 2007;27(Suppl. 2):S303–S307.

[22] ICH. Q6B Specifications: Test Procedures and Acceptance Criteria for Biotechnological/ Biological Products, 1999. http://www.fda.gov/cder/guidance/Q6Bfnl.PDF.

[23] Rosenberg AS. Immunogenicity of biological therapeutics: a hierarchy of concerns. *Dev Biol* 2003;112:15–21.

[24] Shankar G, Shores E, Wagner C, Mire-Sluis A. Scientific and regulatory considerations on the immunogenicity of biologics. *Trends Biotechnol* 2006;24:274–280.

[25] Shankar G, Pendley C, Stein KE. A risk-based bioanalytical strategy for the assessment of antibody immune responses against biological drugs. *Nat Biotechnol* 2007;25:555–561.

[26] Kozlowski S, Swann P. Current and future issues in the manufacturing and development of monoclonal antibodies. *Adv Drug Deliv Rev* 2006;58:707–722.

[27] Schenerman MA. Defining critical quality attributes and linking to QbD and molecular structure. In: Rathore AS, Mhatre R, editors. Quality by Design for Biopharmaceuticals: Perspectives and Case Studies. Wiley Interscience; 2009.

[28] Kourti T. Process analytical technology and multivariate statistical process control: wellness index of product and process—part 3. *J Process Anal Technol* 2006;3:18–24.

[29] PDA. Technical Report 42, (TR42) Process Validation of Protein Manufacturing, 2005.

[30] Senger RS, Karim MN. Effect of shear stress on intrinsic CHO culture state and glycosylation of recombinant tissue-type plasminogen activator protein. *Biotechnol Prog* 2003;19:1199–1209.

[31] Wong DC, Wong KT, Nissom PM, Heng CK, Yap MG. Targeting early apoptotic genes in batch and fed-batch CHO cell cultures. *Biotechnol Bioeng* 2006;95:350–361.

[32] Delvigne F, Lejeune A, Destain J, Thonart P. Stochastic models to study the impact of mixing on a fed-batch culture of *Saccharomyces cerevisiae*. *Biotechnol Prog* 2006;22:259–269.

[33] Kourti T. The process analytical technology initiative and multivariate process analysis, monitoring and control. *Anal Bioanal Chem* 2006;384:1043–1048.

[34] U.S. FDA. Frequently Asked Questions About Therapeutic Biological Products: Question 10, 2006. http://www.fda.gov/cder/biologics/qa.htm.

[35] U.S. FDA. FDA Guidance Concerning Demonstration of Comparability of Human Biological Products, Including Therapeutic Biotechnology-Derived Products, 1996. http://www.fda.gov/cder/guidance/compare.htm.

[36] Eriksson E, Johansson E, Kattaneh-Wold N, Wikstrom C, Wold S. Design of Experiments, Principles and Applications. Umea, Sweden: Umetrics Academy; 2000.

[37] Lewis GA, Mathieu D, Phan-Tan-Luu R. Pharmaceutical Experimental Design. New York: Informa Healthcare; 2007.

[38] McDermott RE, Mikulak RJ, Beauregard MR. The Basics of FMEA. New York: Productivity Press; 1996.

[39] Stamatis DH. Failure Mode and Effects Analysis: FMEA from Theory to Execution. 2nd ed. Milwaukee, WI: ASQ Quality Press; 2003.

[40] Seely JF, Seely RJ. A rational step-wise approach to process characterization. *BioPharm Int* 2003;16:24–34.

[41] Kelley BD. Establishing process robustness using designed experiments. In: Sofer G, Zabriskie DW, editors. Biopharmaceutical Process Validation. New York: Marcel Dekker; 2000. p 29–59.

[42] Kelley BD, Jennings P, Wright R, Briasco C. Demonstrating process robustness for chromatographic purification of a recombinant protein. *BioPharm Int* 1997;10:36–47.

[43] U.S. FDA. Guidance for Industry: Quality Systems Approach to Pharmaceutical CGMP Regulations, 2006.

[44] ICH. Q10 Pharmaceutical Quality System Step 2, 2007. (step 4, 6/4/2008, http://www.ich.org/cache/compo/363-272-1.html#Q10).

[45] Johanson JR. A rolling theory for granular solids. *J Appl Mech* 1965;32:842–848.

[46] Roberts RJ, Rowe RC, York P. The relationship between the fracture properties, tensile strength and critical stress intensity factor of organic solids and their molecular structure. *Int J Pharm* 1995;125:157–162.

[47] Chen J, Yang T, Cramer SM. Prediction of protein retention times in gradient hydrophobic interaction chromatographic systems. *J Chromatogr A* 2008;1177:207–214.

[48] ICH. M4Q: The CTD—Quality, 2001. http://www.fda.gov/cder/guidance/4539q.pdf.

[49] U.S. FDA. Pharmaceutical Quality for the 21st Century: A Risk-Based Approach Progress Report, 2007. http://69.20.19.211/oc/cgmp/report0507.html#appendix17.

[50] EFPIA. Mock P2 for "Examplain" Hydrochloride, 2006. http://www.efpia.org/content/default.asp?PageID=263&DocID=2933.

[51] Chen C. CMC Pilot Program: What are the Best Practices for Quality Assessment: an FDA Perspective. Roundtable on Challenges and Opportunities for Implementation of Quality by Design (QbD). Meeting, AA. San Diego, CA; 2007.

[52] U.S. FDA. Minutes for October 5, 2006 Meeting, Advisory Committee for Pharmaceutical Science, 2006.

[53] Winkle HN. Implementing Quality by Design. Evolution of the Global Regulatory Environment: A Practical Approach to Change, 2007. http://www.fda.gov/Cder/OPS/ImplementingQualitybyDesign.pdf.

[54] CFR. Title 21 Changes to an approved application: Protocols. 601.12(e), 2006.

[55] CFR. Title 21 Supplements and other changes to an approved application: Protocols. 314.70 (e), 2006.

[56] Kozlowski S. Protein therapeutics and the regulation of quality: a brief history from an OBP perspective. *BioPharm Int* 2007;20(10):37–55.

3

MOLECULAR DESIGN OF RECOMBINANT MALARIA VACCINES EXPRESSED BY *Pichia pastoris*

David L. Narum

3.1 INTRODUCTION

Malaria continues to threaten human health and economic development within tropical countries. Malaria-related death rates have doubled over the past 30 years with an annual death rate of 1–2 million people, mostly children younger than 5 years of age from sub-Saharan Africa [1]. More than 40% of the world's population is at risk of getting infected. Malaria is a disease caused by the infection of a protozoan parasite from the genus *Plasmodium*, which is transmitted by female *Anopheles* mosquitoes. The most severe form of the disease is caused by *Plasmodium falciparum*. Common clinical manifestations of malaria due to the presence of parasites infecting erythrocytes are episodes of chills, fever, and sweating. Other symptoms include headache, malaise, fatigue, and body aches. Depending on the immunological status of the individual, infection can lead to severe anemia, kidney failure, or cerebral malaria (coma), and ultimately to death.

Current efforts to control malaria include use of drugs for treatment of the clinical disease as well as insecticides for treating human dwellings and more recently bednets. The use of insecticide-treated bednets has had a significant impact on reducing child mortality [2]. The widespread use of antimalarial drugs has led to the development of significant drug resistance, which has limited control efforts. Current treatment regimens

Quality by Design for Biopharmaceuticals, Edited by A. S. Rathore and R. Mhatre
Copyright © 2009 John Wiley & Sons, Inc.

indicate the use of combination therapies, potentially making drug delivery more expensive. Due to insecticide resistance by the mosquito and drug resistance by the parasite, it is generally accepted that the development of a malaria vaccine is critical to future efforts for the control (or possible elimination) of malaria.

The malaria parasite has a complex life cycle, which involves release of the parasite (sporozoite) while probing for a blood vessel into the individual's skin before a blood meal. The parasite travels to and develops within a liver hepatocyte, where it eventually ruptures and releases parasites (merozoites) that cyclically invade and develop within erythrocytes. Subsets of the parasitized erythrocytes develop into male and female parasites (gametocytes) that are subsequently taken up by a mosquito during a blood meal. Following this blood meal, fertilization occurs in the midgut of the mosquito and the parasite then migrates through the mosquito midgut to replicate beneath the midgut wall. The complexity of the malaria parasite life cycle, in which each stage is immuno-logically distinct, has led to diverse approaches for vaccine development. The diverse approaches include various stage-specific vaccine targets: sporozoite and liver stages or a "pre-erythrocytic" vaccine; blood-stage or an "erythrocytic" vaccine, which would include a pregnancy vaccine; and a transmission blocking vaccine. These stage-specific vaccines would in principle provide sterile immunity, disease control, or reduced malaria transmission to humans. In conjunction with numerous stage-specific parasite targets, there are numerous vaccine delivery systems being studied, including DNA vaccines, attenuated sporozoite vaccines, viral-vectored vaccines, virus-like particle vaccines, recombinant protein vaccines, or combinations of these different delivery systems (for a summary of the above, see Ref. [3]). To date, the only malaria vaccine that has shown some clinical efficacy is RTS/S, a pre-erythrocytic virus-like particle vaccine [4].

In the mid-1990s, the development of recombinant protein vaccines for malaria was limited, in principle, to recombinant expression in *Escherichia coli* or baculovirus. *E. coli* production necessitated the development of refold procedures for provision of relevant protein structure while baculovirus production was limited by scalability of fermentation and multiplicity of viral infection rates. Therefore, we (members of the Malaria Group at EntreMed, Inc.) as well as others [5, 6] were interested to develop the methylotrophic yeast *Pichia pastoris*, a eukaryotic expression system, for scalable production of correctly folded recombinant malarial proteins from *P. falciparum* for human clinical trials. However, based on initial experimentation, it became apparent that a systematic approach would be needed to achieve this result.

Quality by Design (QbD) is a planned approach to pharmaceutical development that may or may not use prior knowledge for planning purposes [7, 8]. QbD is currently being applied to development of new pharmaceutical products and in particular to the development of therapeutic monoclonal antibodies [9, 10]. In this chapter, the progres-sion from preclinical development through pilot-scale production of several malarial proteins following a QbD approach for development of a pharmaceutical drug substance will be discussed. In the case studies described here, the following will be presented: (1) use of a QbD approach to design a synthetic gene and its gene product (drug substance) with specified critical quality attributes (CQAs) for production using the *P. pastoris* expression system; (2) a comparison of codon usage in gene design for *P. pastoris* expression of a malarial protein; and (3) impact of modifying the expression host by overexpressing a chaperone identified as protein disulfide isomerase (PDI). These approaches have facilitated the development of conformance parameters in which critical process parameters and CQAs of the bulk drug substances were preset, including parameters such as protein structure, glycosylation, functional activity, biochemical characteristics, and so on (for a complete summary of important parameters, see Table 3.2). The manufacturability at pilot scale was also established such that the production levels provided sufficient quantities of a quality bulk drug substance. The process parameters for production were divided into four parts: molecular design of the gene and gene product, fermentation, recovery of a secreted material, and purification. CQAs of the bulk drug substance were evaluated throughout the production process to understand and characterize the functional relationships between process parameters and drug substance, which is part of a QbD approach [7, 8].

For the purpose of this chapter, three malarial proteins will be discussed with regard to their development following a QbD approach. Two erythrocytic *P. falciparum* merozoite proteins, the erythrocyte binding antigen-175 (EBA-175) and the apical membrane antigen 1 (AMA1), will be discussed with regard to their expression in *P. pastoris* and subsequent characterization of the bulk substances. A third recombinant malarial protein, Pfs25, which is a 25 kDa transmission blocking vaccine target [11], will be discussed briefly regarding its expression and biochemical characterization when influenced by co-overexpression of PDI [12]. Both the blood-stage malaria proteins are involved in merozoite invasion of erythrocytes based on *in vitro* growth studies and *in vivo* protection studies [13, 14]. EBA-175, or more specifically a region identified as region II (RII) (Fig. 3.1), binds sialic acid residues on its receptor glycophorin A, in conjunction with the peptide backbone on erythrocytes [15]. Antibodies generated against RII inhibit parasite invasion *in vitro* [16] and provide protection against a virulent

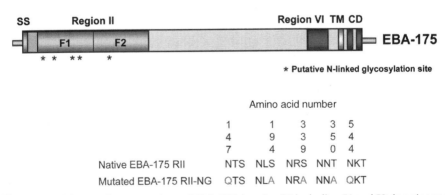

Figure 3.1. Scheme of gene structure of EBA-175 showing RII including F1 and F2 domains and the division of six other regions (upper panel). The lower panel shows the native EBA-175 RII amino acid sequences in which the putative N-linked glycosylation sites are identified (NxS/T) by amino acid position using the sequence identified by accession number XP_001349207. The five amino acid substitutions used for expression of EBA-175 RII-NG are shown in gray. The position of putative N-linked glycosylation sites are noted by asterisks. CD, cytoplasmic domain; SS, signal sequence; TM, transmembrane domain.

challenge infection in a nonhuman primate model [17, 18]. Although the specific function of AMA1 is unknown and still under investigation, it is believed to have an erythrocyte binding function based on different assays [14, 19, 20]. At this time all three proteins (EBA-175 RII, AMA1, and Pfs25) have been cGMP manufactured, at pilot scale, for human clinical trials (AMA-1 [21, 22]; Pfs25 [23]; EBA-175 RII-NG, Hall et al., unpublished). The QbD approach for the molecular development and successful expression in *P. pastoris* of each of these three recombinant proteins will be discussed.

3.2   THE MALARIA GENOME AND PROTEOME

The proteome of *P. falciparum* is estimated to encode more than 5000 proteins [24], of which approximately 5% or 250 proteins may be of interest for vaccine development. An analysis of the literature would suggest that approximately 100 malarial proteins have been recombinantly expressed in some form and analyzed to a limited extent. The number of malarial proteins under investigation for *P. falciparum* vaccine development using the various vaccine delivery systems described above is 47 candidates, including 31 in preclinical development and 16 in clinical development [3]. Of the 16 in clinical development, the number of recombinant-protein-based vaccines is about 10 due in part to the difficulty in expressing malarial proteins with relevant structure. Part of the difficulty in expressing malarial proteins in recombinant expression systems is due to the unique content of the genome of *P. falciparum*. Analysis of the *P. falciparum* genome shows that the A + T content is 76.2%, compared to 57.3% in *P. pastoris* and to 47.7% in humans (*Homo sapiens*) (http://www.kausa.or.jp/codon/). The A + T genomic bias of *P. falciparum* negatively impacted expression of EBA-175 RII in *P. pastoris* using the native gene. Efforts to produce EBA-175 RII in *P. pastoris* using the native A + T-rich gene failed apparently due to premature termination of mRNA transcripts as observed by Northern blot [25]. Similar observations were reported for AMA1 FVO [5] and the human immunodeficiency virus type 1 envelope glycoprotein, gp120 [26], when expressed in *P. pastoris* using native genes. The native gp120 gene sequence was identified to contain a *Saccharomyces cerevisiae* polyadenylation consensus sequence that acted as a premature polyadenylation site, resulting in the production of a truncated mRNA transcript [26]. In the case of EBA-175 RII, stretches of poly A's may be observed within the native sequence such that these may act as premature polyadenylation sites (Fig. 3.2).

3.3   EXPRESSION OF TWO MALARIA ANTIGENS IN *P. pastoris*

In two independent preclinical development efforts, EBA-175 RII and AMA1 were shown to be malaria vaccine candidates. Analyses of baculovirus-expressed proteins demonstrated for each that antigen-specific antibodies inhibited parasite development *in vitro* [16, 27], *in vivo*, or both [17]. Despite an interest in pursuing human clinical studies, progression was hindered due to a bottleneck in identifying a manufacturing process for the production of these two proteins in a scalable manner with suitable

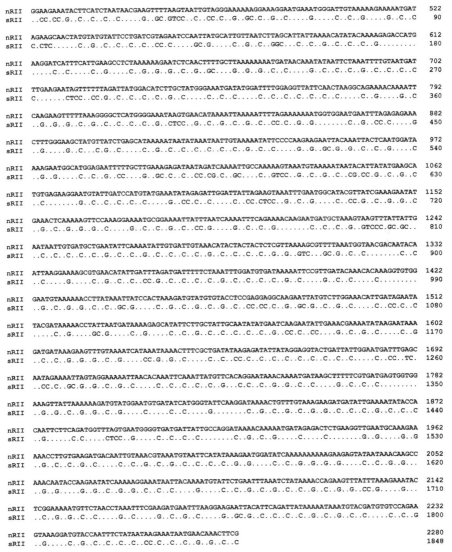

Figure 3.2. Nucleotide sequence alignment for native *P. falciparum* 3D7 EBA-175 RII (nRII) and synthetic mammalian codon optimized RII-NG (sRII).

quantity and quality attributes. At the time, the baculovirus expression system was not considered suitable for scalable production, for reasons stated above. The *P. pastoris* expression system was of particular interest due to its scalability, capacity for posttranslational modifications (higher probability of correct disulfide bond formation), and ease of purification of a secreted product.

In the late 1990s, expression projects were being developed around the use of highly expressed genes with bias toward codons that are recognized by the most abundant tRNA

TABLE 3.1. Codon Usage for Most Abundant Amino Acid for *P. falciparum, H. sapiens,* and *P. pastoris*

Amino Acid	*P. falciparum*	*H. sapiens*	*P. pastoris*	Amino Acid	*P. falciparum*	*H. sapiens*	*P. pastoris*
A	GCA	GCC	GCU	L	UUA	CUG	UUG
R	AGA	CGG	AGA	K	AAA	AAG	AAG
N	AAU	AAC	AAU	M	AUG	AUG	AUG
D	GAU	GAC	GAU	F	UUU	UUC	UUU
C	UGU	UGC	UGU	P	CCA	CCC	CCA
Q	CAA	CAG	CAA	S	AGU	AGC	UCU
E	GAA	GAG	GAA	T	ACA	ACC	ACU
G	GGU	GGC	GGU	W	UGG	UGG	UGG
H	CAU	CAC	CAU	Y	UAU	UAC	UAC
I	AUU	AUC	AUU	V	GUA	GUG	GUU

Nucleotide that varies with respect to *P. falciparum* is underlined.

species in the organism [28]. Depending on the organism, each amino acid, with the exception of methionine and tryptophan, can be encoded by two–six different synonymous codons. The frequencies at which these synonymous codons are used depend on the level of protein expression and differ among organisms (Table 3.1). The Malaria Group at EntreMed, Inc. previously reported that a mammalian codon optimized synthetic gene encoding EBA-175 RII expressed by a DNA vaccine plasmid increased EBA-175 RII protein expression levels *in vitro* and immunogenicity levels *in vivo* [29]. The mammalian codon optimized EBA-175 RII gene lacked the stretches of poly A's which were responsible for the failed expression in *P. pastoris*. Thus, at the same time, a parallel development effort was ongoing to evaluate whether the synthetic EBA-175 RII gene would overcome the problem of early transcription termination events, which had been observed in *P. pastoris* with the native gene, and thereby express full-length RII protein. A similar approach was being used by two other groups (The Malaria Vaccine Development Branch (MVDB), National Institute of Allergy and Infectious Disease, National Institute of Health, Rockville, Maryland, and The Biomedical Primate Research Centre, Rijswijk, The Netherlands), along with EntreMed, Inc. for the production of *P. falciparum* AMA1 in *P. pastoris* [5, 30].

Each group independently aimed at developing an approach that would allow the manufacturing of recombinant malarial antigens, which would maintain CQAs of the bulk drug substances that were previously observed for the two recombinant baculovirus proteins (EBA-175 RII [31] and AMA1 [32]). In addition, it was necessary to identify a production process that was suitable for manufacturing a sufficient quantity of bulk product using a robust process. The quality of the bulk drug substance was also important with regard to its integrity, purity, functional activity, and so on (see Table 3.2 for a complete list). In the following sections, the molecular (gene) design for recombinant expression of these two malarial proteins in *P. pastoris* is presented. Limited characterization of the CQAs for the EBA-175 RII bulk drug substance is presented, which justify the QbD approach used here that lead to its cGMP pilot-scale production. Regarding

TABLE 3.2. Potential Critical Process Parameters and Quality Attributes Evaluated During Development

Potential critical process parameters for product development	
Gene design/codon usage	Osmolarity
Gene boundaries	Quantity
Secretion of product into fermentation media	Stability
Induction conditions during fermentation:	Integrity or extent of "proteolytic nicking" during fermentation
Ph	
Temperature	
Methanol feed rate	
Potential CQAs and analytical method(s) for evaluation	
Integrity (reduced and nonreduced)	Structure, conformation, or function
SDS-PAGE	Immunoblot with conformational mAbs
RP-HPLC	Plasmon surface resonance
	Circular dichroism spectrum
	Functional binding assay
Purity (reduced and nonreduced)	Posttranslational modifications
SDS-PAGE	N-linked glycosylation
RP-HPLC	O-linked glycosylation
	Other
Solubility (SEC-HPLC)	N-terminal sequence
Aggregation	Peptide mapping (reduced and nonreduced)
MALS-SEC-HPLC	
Sedimentation velocity or equilibrium	Stability
Mass	Impurity profile
Matrix-assisted laser desorption ionization (MALDI) spectroscopy	Host cell protein content
	Endotoxin
Electron spray ionization spectroscopy	DNA content

MALS-SEC-HPLC, multiangle light scattering-SEC-HPLC; RP-HPLC, reverse-phase high-performance liquid chromatography; SDS-PAGE, sodium dodecyl sulfate-polyacrylamide gel electrophoresis; SEC-HPLC, size exclusion chromatography-HPLC.

AMA1, the impact of codon usage for generating a synthetic gene for expression in *P. pastoris* will be discussed, which lead to its successful cGMP pilot-scale production [30].

3.3.1 Production of Recombinant EBA-175 RII Protein

***P. pastoris* Expression of EBA-175 RII.** A comparison of the native EBA-175 RII gene sequence and the synthetic-mammalian-encoded EBA-175 RII gene sequence is shown in Fig. 3.2. A total of 591 out of 1848 nucleotides (32.0%) were altered in the synthetic RII gene. The result was that the A +T content of the RII gene changed from

Coomassie blue-stained SDS-PAGE gel

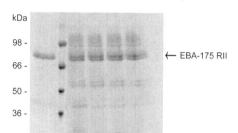

Figure 3.3. Time course of *P. pastoris* expression and secretion of recombinant EBA-175 RII by shake flask fermentation. Supernatants were characterized by SDS-PAGE 4–20% gradient gel analysis using Coomassie blue stain under nonreducing conditions. Doublet primarily represents expression of N-linked and non-N-linked EBA-175 RII proteins. Control protein was purified baculovirus-expressed EBA-175 RII [31].

72.7% to 43.3% and the potential internal polyadenylation sites were removed. The synthetic gene encoding EBA-175 RII was cloned into the pPICZαA vector and transformed in *P. pastoris* host strain X33 following standard procedures [33]. The pPICZαA vector includes the α-factor secretion signal from *S. cerevisiae* for directing secreted expression of the recombinant protein. Transformants were screened by shake flask fermentation using methanol as the sole carbon source for induction of the alcohol oxidase I gene promoter. The results of shake flask fermentation for one clone are shown in Fig. 3.3. *P. pastoris* expressed EBA-175 RII as a secreted product, which was observed as a doublet; the lower band migrated at the expected molecular mass as determined by Coomassie blue-stained SDS-PAGE gel analysis. For preclinical studies, recombinant EBA-175 RII protein was purified by a series of column chromatography steps including cation exchange, mixed mode hydrophobic interaction, and a second cation exchange column [33]. An example of the purified recombinant EBA-175 RII protein is shown in Fig. 3.4. The identity of the recombinant protein was determined by other biochemical and biophysical techniques, including erythrocyte binding assay (Fig. 3.6), and immunoblot with conformation-dependent monoclonal antibodies [34].

EBA-175 RII protein contains five putative N-linked glycosylation sites based on the conservation of the amino acid sequence NxS/T. The locations of the putative sites are shown in Figs 3.1 and 3.5. Analysis of the purified recombinant EBA-175 RII protein by carbohydrate analysis demonstrated the presence of N-linked carbohydrates, that is, 20% or one in five sites randomly appeared to contain *N*-linked glycosylation with $MAN_{14-16}GlcNAc_2$ [33]. The native EBA-175 RII domain does not contain N-linked glycosylated moieties due to the inability of *P. falciparum* parasites to N-glycosylate proteins [35]. The presence of the *N*-linked carbohydrates could alter the immunogenicity of the recombinant protein. Such an observation had been made for another *P. falciparum* blood-stage protein identified as the 42 kDa fragment of the

Coomassie blue-stained
SDS-PAGE gels

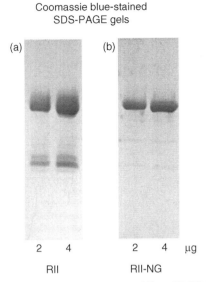

Figure 3.4. Purified *P. pastoris*-expressed EBA-175 RII (A) or RII-NG (B) protein as detected by Coomassie blue-stained SDS-PAGE gel analysis. Purified recombinant proteins were separated using a 4–20% gradient gel under nonreducing conditions.

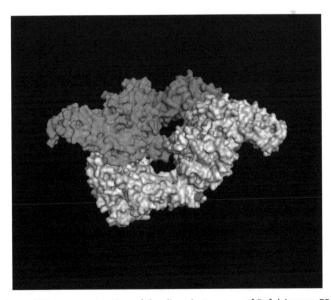

Figure 3.5. Space-filling representation of the dimeric structure of *P. falciparum* EBA-175 RII [47] showing positions of amino acid substitutions for expression of RII-NG. Monomers are shaded in cyanoblue and gray. The four amino acid substitutions made, which can be identified in the crystal structure, are shown in red. (See the insert for color representation of this figure.)

merozoite surface protein 1 ($MSP1_{42}$) [36]. Recombinant forms of $MSP1_{42}$ protein were expressed in transgenic mice either with or without a putative N-linked glycosylation site present near a critical enzymatic cleavage site [37]. These two recombinant forms of $MSP1_{42}$ were purified and tested for their capacity to induce immunological protection against a blood-stage challenge infection using a primate malaria model. Despite both forms having a similar immunogenicity profile, the recombinant N-linked glycosylated $MSP1_{42}$ protein failed to induce any protective response compared to the recombinant non-N-linked glycosylated protein [36].

The observation that the recombinant N-linked glycosylated $MSP1_{42}$ protein failed to protect *Aotus* monkeys against a challenge infection impacted the malaria vaccine community in such a manner that it became a necessity to remove putative N-linked glycosylation sites from recombinantly expressed proteins derived from eukaryotic expression systems such as *P. pastoris*. This impacted the QbD approach for the development of EBA-175 RII in one important manner. It provided for the improvement of a CQA by eliminating the presence of N-linked carbohydrates, which subsequently lead to an increase in the quantity of the bulk drug substance. However, it also added a level of uncertainty to the integrity of the recombinant protein's structure, which could alter its immunogenicity. In the case of EBA-175 RII, it was decided to mutate the putative N-linked glycosylation sites and characterize the recombinantly expressed protein for its functional activity, that is, binding erythrocytes (given subsequently) and induction of inhibitory antibodies.

P. pastoris Expression and Characterization of Non-N-Glycosylated EBA-175 RII (RII-NG).

EBA-175 RII was expressed and purified as a non-N-glycosylated protein (identified as RII-NG) using the same expression plasmid, shake flask fermentation, and purification strategy, already described. The mammalian-encoded synthetic gene was used as a template. The five putative N-linked glycosylation sites identified within EBA-175 RII were mutated to remove the consensus sequence NxS/T. The following hierarchy was used to introduce the point mutations. First, homologous deduced amino acid sequence alignments from other *P. falciparum* strains expressing EBA-175 were evaluated for naturally occurring point mutations. Second, paralogous [38–40] and orthologous [41] deduced amino acid sequence alignments were evaluated for possible amino acid substitutions. Finally, a "best guess" was used such that asparagine is mutated to glutamine and serine/threonine is mutated to alanine. The amino acid sequence positions are shown in Fig. 3.1. Even though the strategy described above was used, none of the alignments provided any help in the selection of the "best" amino acid for substitution. Thus, the gene was mutated such that either the first amino acid (N–Q) or third amino acid (S/T–A) was substituted to remove the consensus sequence NxS/T. Transformants were screened by shake flask fermentation and analyzed by Coomassie blue-stained SDS-PAGE gel. A production clone was selected and used to produce purified recombinant EBA-175 RII-NG protein. A Coomassie blue-stained SDS-PAGE gel showing a comparison of EBA-175 RII and RII-NG is shown in Fig. 3.4. Except for the absence of N-linked glycosylation, the biophysical characterization of purified EBA-175 RII-NG demonstrated that it appeared similar to EBA-175 RII protein, as determined by immunoblot with conformation-dependent mAbs [34] and

Human RBC binding assay

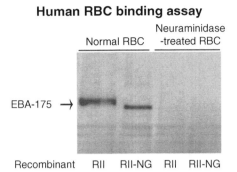

Figure 3.6. Immunoblot showing binding specificity of EBA-175 RII and RII-NG proteins. Recombinant region II was allowed to bind RBC or neuraminidase-treated RBC that had sialic acids removed. The mixture was spun through oil-separating RBC with bound or unbound RII or RII-NG protein. The RBC bound RII or RII-NG protein was eluted off RBC with high salt buffer. Recombinant RII or RII-NG bound normal RBC but not neuraminidase-treated RBC. The immunoblot was probed with rabbit anti-RII antibodies.

sialic-acid-dependent binding to human erythrocytes (Fig. 3.6). Therefore, the CQA identified as maintaining functional activity was unchanged.

Structural Comparison of EBA-175 RII and RII-NG. Purified EBA-175 RII-NG protein was used for crystallographic studies and shown to crystallize as a dimer [42]. For the purpose of this discussion, a space-filling representation of the dimeric structure is shown in Fig. 3.5. Four of the five mutations are shown in red and can be observed together when viewed as a dimer. The fifth mutation is near the amino terminus of the recombinant protein and is not represented in the crystal structure. No significant changes were observed due to the introduction of the amino acid substitutions.

Comparison of EBA-175 RII and RII-NG Immunogenicity and Parasite Growth Inhibition. To evaluate the immunogenicity of EBA-175 RII and RII-NG, rabbits were immunized with both recombinant proteins [33]. The results of the ELISA titers as detected by a homologous and heterologous capture antigen are shown in Table 3.3. No marked differences were observed by ELISA. Rabbit immunoglobulin G was purified from the immune sera and tested in a *P. falciparum* growth inhibition assay, which generally measures the capacity of antibodies to interfere with erythrocyte invasion. The level of growth inhibition appeared similar for the two different recombinant forms of EBA-175 when measured against two different strains of *P. falciparum*, which may use different erythrocyte invasion pathways [16]. Again, another important CQA appeared to be unchanged.

Human Clinical Testing. EBA-175 RII-NG has been cGMP manufactured for human clinical testing by a contract facility. The recombinant RII-NG protein is formulated on aluminum phosphate [43] and is currently being tested in a phase I dose escalation safety study. The results of this phase I human trial may be available in 2009.

TABLE 3.3. Antibodies Against Recombinant RII and RII-NG Proteins Block Function of EBA-175

| Rabbit No. | Recombinant P. pastoris RII Vaccine[c] | RII ELISA Titer[a] | | % Growth Inhibition of Erythrocyte Invasion[b] | |
		RII Capture Antigen	RII-NG Capture Antigen	FVO Parasites	3D7 Parasites
74	RII	937,296	623,374	22	25
75		1,656,000	1,442,000	29	ND
76	RII-NG	355,233	562,888	20	23
77		244,534	407,154	18	17

[a]Rabbits 74/75, 76/77, and controls were immunized thrice with 100 μg recombinant protein in Freund's adjuvant or adjuvant alone for controls.
[b]Interpolated reciprocal RII ELISA titers are reported for an OD of 0.5 using either RII or RII-NG for the capture antigen.
[c]%Growth inhibition reported using purified rabbit IgG at 1.0 mg/mL compared to control. All values were statistically significant $p < 0.05$.

3.3.2 *P. pastoris* Expression of Apical Membrane Antigen 1 (AMA1): Codon Selection

In the following example, the selection of the codon usage is discussed for another important malarial vaccine target, that is, recombinant AMA1 3D7 protein. A molecular schematic of the protein is shown in Fig. 3.7. Two different approaches for the molecular design of the synthetic AMA1 3D7 gene were used. The Malaria Group from EntreMed, Inc. selected to use a mammalian codon optimized gene while MVDB selected to use a *P. pastoris* codon optimized gene for expression in *P. pastoris*. A codon usage table similar to that used by the two groups is shown in Table 3.1, which compares the optimum codons used by *P. falciparum* to that of *H. sapiens* (mammalian) or *P. pastoris*. A comparison of the native, mammalian codon optimized and *P. pastoris* codon optimized genes is shown in Fig. 3.8. A total of 525 out of 1560 nucleotides (33.7%) were altered in the synthetic mammalian gene; a total of 306 out of 1575 nucleotides (19.4%) were altered in the *P. pastoris*-encoded gene. The A +T content of the AMA1 genes changed from 69.4% for the native gene to 38.0% for the mammalian and 58.8% for *P. pastoris* genes. Most of the nucleotide changes occurred in the third base position. Most striking is that the long stretches of poly A's were removed to prevent the formation of truncated mRNA transcripts. Six putative N-linked glycosylation sites were identified and their amino acid sequence and position are shown in Fig. 3.7. In both of these synthetic genes, the codons were changed to remove the putative N-linked glycosylation sites using a similar methodology to that discussed above for EBA-175 RII. Appropriate restriction enzyme sites were modified to allow cloning in either pPICZαA (Narum, Merchant, Fuhrmann, and Kumar, unpublished) or pPIC9K [17]. Expression plasmids were

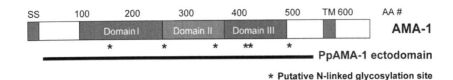

	Amino acid number				
	1 6 2	2 8 6	3 7 1	4 4 2 2 1 2	4 9 9
Native AMA-1	NTT	NYT	NAS	NNSS	NST
Syn *P. pastoris* AMA-1	KTT	NLV	NRE	NDKN	QST
Syn mammalian AMA-1	KTT	NYV	NAA	NQAS	QST

<u>Figure 3.7.</u> Scheme of gene structure of AMA1 showing division of domains I–III and other domains (upper panel). The lower panel shows the native AMA1 3D7 amino acid sequences in which the putative N-linked glycosylation sites are identified (NxS/T) by amino acid position using the sequence identified by accession number: U65407. The six amino acid substitutions used for expression of the synthetic *P. pastoris* or mammalian codon optimized gene are shown in red. The solid black bar represents the region recombinantly expressed in *P. pastoris*. The position of putative N-linked glycosylation sites are noted by asterisks. CD, cytoplasmic domain; SS, signal sequence; TM, transmembrane domain. (See the insert for color representation of this figure.)

linearized and appropriate strains (X33 or GS115) of *P. pastoris* were transformed and placed under selection pressure. Recombinant AMA1 protein production clones were identified using either the synthetic mammalian codon optimized gene (Narum, Merchant, Fuhrmann, and Chen, unpublished) or *P. pastoris* codon optimized gene [17].

Recombinant AMA1 3D7 protein was produced by both groups. Recombinant AMA1 3D7 protein produced by MDVB [17] was extensively characterized using most of the analytical methods shown in Table 3.2. For the purpose of this discussion, there are two important CQAs that were observed in both the recombinant AMA1 3D7 proteins, even with the mutations to remove the six putative N-linked glycosylation sites (Fig. 3.7). First, both recombinant AMA1 3D7 proteins induced growth inhibitory antibodies as detected by inhibition of parasite growth *in vitro* (Narum and Haynes, unpublished and Ref. [17]). Second, both recombinant proteins maintained a conserved structural epitope as determined by immunoblotting with a conformation-dependent monoclonal antibody identified as 4G2 (data not shown).

To conclude, expression of numerous malarial proteins by *P. pastoris* has been made possible by changing the codon usage of the native gene that prevents premature termination of mRNA transcription. The significance of which optimized codons to use, mammalian, *P. pastoris*, or other, must be empirically tested. In addition, there are other factors, which are not discussed here, that may be important to consider such as the expression vector or promoter being used, potential for formation of mRNA hairpins, as

```
nPfAMA1  GGAAGAGGACAGAATTATTGGGAACATCCATATCAAAATAGTGATGTGTATCGTCCAATCAACGAACATAGGGAACATCCAAAAGAATAC  153
syPfAMA1 ---TAC.T......C..C...............C.....CTC......CA.G.....A..T......C.T..........G....T        87
smPfAMA1 ..CC.C..C.....C..C....G..C..C..C..G..CTCC..C....C..C..C...........G..CC.C..G..C..C..G..G...   90

nPfAMA1  GAATATCCATTACACCAGGAACATACATACCAACAAGAAGATTCAGGAGAAGACGAAAATACATTACAACACGCATATCCAATAGACCAC  243
syPfAMA1 .....C......G...............T..T...............C..G...........C..T..G...........C.....T.....  177
smPfAMA1 ..G..C..CC.G........G..C..C....G..G..G..C..C..C..G.....G..C..CC.G..G.....C..C..C..C..C.....   180

nPfAMA1  GAAGGTGCCGAACCCGCACCACAAGAACAAAATTTATTTTCAAGCATTGAAATAGTAGAAAGAAGTAATTATATGGGTAATCCATGGACG  333
syPfAMA1 .....A..T................................C..G.....C.....T..C.....TC...C..C.....A..C.........  267
smPfAMA1 ..G..C.....G......C..C..G..G..G..CC.G..C..CTC...C..G..C..G..GC.CTCC..C..C.....C..C..C.....C   270

nPfAMA1  GAATATATGGCAAAATATGATATTGAAGAAGTTCATGGTTCAGGTATAAGAGTAGATTTAGGAGAAGATGCTGAAGTAGCTGGAACTCAA  423
syPfAMA1 ..C........G..C.....C..............C..A..T.....C.....G..G.........G.....C..G..G..A...        357
smPfAMA1 ..G..C......C..G..C..C..C..G..G..C..C..C..CC.C..G..CC.G..C..G..C..G..C..C..G..G..C..C..C..G   360

nPfAMA1  TATAGACTTCCATCAGGGAAATGTCCAGTATTTGGTAAAGGTATAATTATTGAGAATTCAAATACTACTTTTTTAACACCGGTAGCTACG  513
syPfAMA1 ..C...........C.T..G......A..G..A..T..C..C.....C..C..G..A..A.....G..T....G...        447
smPfAMA1 ..CC.C..G..C..C..C..G..C..C..G..C..C..G..C..C..C.....C..C..G..C..C..CC.G..C..C..G..C..C     450

nPfAMA1  GGAAATCAATATTTAAAAGATGGAGGTTTTGCTTTTCCTCCAACAGAACCTCTTATGTCACCAATGACATTAGATGAAATGAGACATTTT  603
syPfAMA1 ..G..C.....C..G..G....G..A....G.........T.........G.....C.....T..G...        537
smPfAMA1 ..C..C..G..CC.G..C..C..C..C..C..C..C..G..C..C..C.....C.......CC.G..C..G..C..C..C..c           540

nPfAMA1  TATAAAGATAATAAATATGTAAAAAAATTTAGATGAATTGACTTTATGTTCAAGACATGCAGGAAATATGATTCCAGATAATGATAAAAAT  693
syPfAMA1 ..C..G...C..G..C..C...C..G..C..G.....C..A..G........G..C.........C.....G..C  627
smPfAMA1 ..C..G..C..C..G..C..G..G....CC.G..C..GC.....CC.G..C..CC.C..C..C.....C..C..C..C..C..G..C      630

nPfAMA1  TCAAATTATAAATATCCAGCTGTTTATGATGACAAAGATAAAAAGTGTCATATATTATATATTGCAGCTCAAGAAAATAATGGTCCTAGA  783
syPfAMA1 ..C..C..C..G..C.........G.....C.........G.....G..A.........T..G..C..C.....G.......C..A...    717
smPfAMA1 ..C..C..C..G..C..C..C..G..C..C..C..G..C..G..C.....CC.G..C..C..C..C..G..G..C..C..C..CC.C     720

nPfAMA1  TATTGTAATAAAGACGAAAGTAAAAGAAACAGCATGTGTTTGTTTTAGACCAGCAAAAGATATATCATTTCAAAACTATACATATTTAAGT  873
syPfAMA1 ..C.....C..G.......TC...G.....T.................G.....T..C.......T.TGGTC..C..GTC.  807
smPfAMA1 ..C..C..C.....GTCC..GC.C...TC.....C..C..CC.C..C..C..G..C..C..C..C..G.....CGTG...CC.GTCC     810

nPfAMA1  AAGAATGTAGTTGATAACTGGGAAAAAGTTTGCCCTAGAAAGAATTTACAGAATGCAAAATTCGGATTATGGGTCGATGGAAATTGTGAA  963
syPfAMA1 ..A..C.......T......G.....C.....G..G.....G..G........A...G..C...  897
smPfAMA1 .....C..G..G..C.....G..G..G.....CC.C.....CC.G.....C..C..G.....CC.G.....G..C..C..C..C..G   900

nPfAMA1  GATATACCACATGTAAATGAATTTCCAGCAATTGATCTTTTTGAATGTAATAAATTAGTTTTTGAATTGAGTGCTTCGGATCAACCTAAA  1053
syPfAMA1 .....T.......C..C..................C...............C..G..G...........C..TC..G........G      987
smPfAMA1 ..C..C..C..G..C..G..C..C.....C..C..G..C..G..C..GC.G..G..C..GC.GC..C..C..C..G..C..G         990

nPfAMA1  CAATATGAACAACATTTAACAGATTATGAAAAAATTAAAGAAGGTTTCAAAAATAAGAACGCTAGTATGATCAAAAGTGCTTTTCTTCCC  1143
syPfAMA1 .....C...........G..T...C..G..C..G.....A.....G..C..A..TAGAGAA.....A..GTC..G........         1077
smPfAMA1 ..G..C..G..G..CC.G..C..C..C..G..C..C..G..G..C.....G..C.........CGCC.......GTCC..C..C..G...    1080

nPfAMA1  ACTGGTGCTTTTAAAGCAGATAGATATAAAAGTCATGGTAAGGGTTATAATTGGGGAAATTATAACACAGAAACACAAAAATGTGAAATT  1233
syPfAMA1 ..A..A..G....G.............C..GTC.....A..A..A..C..C......G..C..T..T..T.....T.......G.......C  1167
smPfAMA1 ..C..C..C..G..C..C.....CC.C..C..GTCC..C..C.....C..C..C.....G..C..G..G..C..G..C..C..G..C      1170

nPfAMA1  TTTAATGTCAAACCAACTGTTTAATTAACAATTCATCATACATTGCTACTACTGCTTTGTCCCATCCCATCGAAGTTGAAAACAATTTT  1323
syPfAMA1 .....C..A..G.....T.....G..C..TG.CAAGAAC..T..C..G..A..A..GC..AGT........A...........T..C..    1257
smPfAMA1 ..C..C..G..G..C..C..CC.G..C..C..C.GG.C..C.....C..C..C..C..CC.......C.......G..G..G......C..C  1260

nPfAMA1  CCATGTTCATTATATAAAGATGAAATAATGAAAGAAATCGAAAGAGAATCAAAACGAATTAAATTAAATGATAATGATGATGAAGGGAAT  1413
syPfAMA1 ........C..G..C..G..........T.....G.....A.........C..G...C..G..C.......C.........T..C       1347
smPfAMA1 ..C..C..CC.G..C..G..C..G..C..G..........GC.C..G..C..G..C..C..GC.G..C..C..C..C..G..C..C      1350

nPfAMA1  AAAAAAATTATAGCTCCAAGAATTTTTATTTCAGATGATAAAGACAGTTTAAAATGCCCATGTGACCCTGAAATGGTAAGTAATAGTACA  1503
syPfAMA1 ..G..G..C..T..G........C.....C..C........G...TC...G..G.........................CTC.C.ATC...T  1437
smPfAMA1 ..G..G..C..C..C..CC.C..C..C..C..C..G..C.......TCCC.G..C.....C..C.....C..G.....GTCCC.GTCC..C   1440

nPfAMA1  TGTCGTTTCTTTGTATGTAAATGTGTAGAAAGAAGGGCAGAAGTAACATCAAATAATGAAGTTGTAGTTAAAGAAGAATATAAAGATGAA  1593
syPfAMA1 ..A.G............C.......G.......C.T.........C..T..C..C..C...............C..G.....         1527
smPfAMA1 ..C..C........C..G..C..G..C..G..GC.CC.C..C..G..G..C..C..C..G..G..G..G..G..G..G...C..G..C..G  1530

nPfAMA1  TATGCAGATATTCCTGAACATAAACCAACTTATGATAAAAATGAAA                                               1638
syPfAMA1 ..C..........C.........G.....A..C.....G.....GACTAGTCACCATCACCATCACCAC                         1596
smPfAMA1 ..C..C..C..C..C..G..C..G..C..C                                                               1560
```

Figure 3.8. Nucleotide sequence alignment comparing the expressed ectodomain of native *P. falciparum* 3D7 AMA1 (nPfAMA1, accession number: U65407) to the synthetic yeast codon optimized (syPfAMA1) and mammalian codon optimized (smPfAMA1) genes. The nucleotide sequences shown here include the relevant codon changes to correspond to the amino acid substitutions identified in Fig. 3.7 (lower panel). The accession number for the syPfAMA1 3D7 amino acid sequence is AF512508.

well as identifying relevant structural boundaries to provide for proper protein folding and stability [44].

3.3.3 Modification of *P. pastoris* Host Strains for Overexpression of Protein Disulfide Isomerase

Another area that has been evaluated to increase recombinant protein production in *P. pastoris* is the augmentation or overexpression of a multifunctional chaperone in the endoplasmic reticulum identified as protein disulfide isomerase. In the *P. pastoris* expression system, recombinant proteins are commonly targeted for secretion to the extracellular media. These proteins are translocated into the endoplasmic reticulum, where folding takes place before secretion through the Golgi apparatus. These proteins must pass a quality control checkpoint before being exported [45]. In the case of a strong expression system like *P. pastoris*, the disulfide exchange machinery may be either overloaded or inefficient in folding high levels of complex proteins. Both cases result in low yield of recombinant protein. Engineering the *P. pastoris* production clone for an increase in both quantity and quality of drug substance adheres to a QbD approach.

PDI is a multifunctional protein that participates in protein folding, assembly, and posttranslational modification in the endoplasmic reticulum. It is capable of functioning both as an enzyme and as a chaperone. PDI catalyzes both the oxidation and isomerization of disulfides of nascent polypeptides. As an enzyme, it increases the rate of disulfide bond formation without altering the folding pathway [46]. While serving as a chaperone, it promotes correct folding of proteins by preventing the misfolding and aggregation of partially folded or misfolded peptides [46]. Elevated levels of PDI activity in bacterial, yeast, and insect cell expression systems increase secretion of heterologous proteins with or without disulfide bonds [47–50].

We evaluated whether overproduction of *P. pastoris* PDI (PpPDI) or a PDI orthologue identified in *P. falciparum* (PfPDI) when overexpressed would increase the production level of a transmission blocking vaccine candidate Pfs25, which contained 11 disulfide bonds and was expressed in *P. pastoris* using a synthetic *P. pastoris* codon optimized gene [51]. The results from this study have been reported and are briefly summarized here [12]. Both PpPDI and PfPDI were cloned into the pPICZαA expression vector using either the native gene or yeast codon optimized gene that was transformed into a *P. pastoris* production clone that expressed the heterologous *P. falciparum* Pfs25 protein. The results of a head-to-head comparison using 5L bioreactors with defined media are shown in Fig. 3.9. Analysis of fermentation supernatant showed the benefit of overexpression of PpPDI and a moderate benefit of the PfPDI (Fig. 3.9). Analysis of the product concentration within the supernatant by high-performance liquid chromatography method demonstrated that overproduction of PpPDI increased the level of Pfs25 protein production by three- to fourfold. Interestingly, not only the quantity of the product was affected but also the quality of the product was significantly improved, as demonstrated by the subsequent analysis of the final bulk protein by electron spray ionization mass spectroscopy. As discussed earlier, malaria proteins lack significant glycosylation [35]. Therefore, the reduction in the level of O-linked mannosylation on Pfs25 due to PpPDI

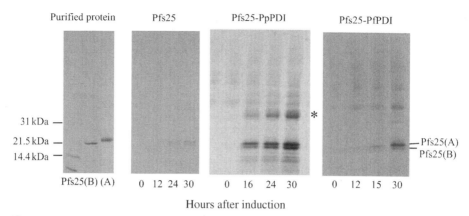

Figure 3.9. Comparison of Pfs25 production by Pfs25, Pfs25-PpPDI, and Pfs25-PfPDI clones. Clarified fermentation supernatants collected at different time points were separated on SDS-PAGE under nonreduced conditions and Coomassie blue stain. Accumulated Pfs25 (A) and (B) recombinant proteins are indicated by small triangles on the right. Purified Pfs25 (A) and (B) (1 μg each) are shown in the right panel as control. Overexpressed PpPDI protein was identified (asterisk) in Pfs25-PpPDI fermentation supernatant. Reprinted with permission from Elsevier [52].

overexpression (Fig. 3.10), which was also confirmed by a monosaccharide composition analysis showing a fivefold reduction in mannosylation [48], provided for a significant increase in quality as well. The benefits of overexpression of PpPDI on the productivity of other *P. pastoris*-expressed malarial proteins have been observed (Narum, unpublished and Ref. [44]).

3.4 SUMMARY

A QbD approach for molecular design has improved the quality and quantity of recombinantly expressed malarial proteins. The capacity to produce malarial proteins using recombinant expression systems has improved significantly with the introduction of synthetic genes for *P. pastoris*. Although previous empirical information played an important part in the early stages of the QbD approach, the targeted molecular design of malarial genes for expression in *P. pastoris* has enabled the development of processes that are suitable for manufacturing at pilot scale (Table 3.4). The *P. pastoris* production clones discussed here all achieved production levels at or above 1000 vaccine doses/L broth, which indicates their suitability for further scale-up. The use of synthetic genes not only enables the expression of a particular protein but also permits appropriate modifications to increase the CQAs related to the quality of the bulk drug substance (Table 3.4). Recombinant EBA-175 RII, AMA1, and Pfs25 were manufactured at pilot scale following cGMP and the drug substances appear to mimic native malarial proteins, as indicated by induction of antiparasite antibodies. Finally, further augmentation in both quantity and quality appear possible by the overexpression of host chaperones such as PDI. These measures require planned development; the result is a quality product by molecular design.

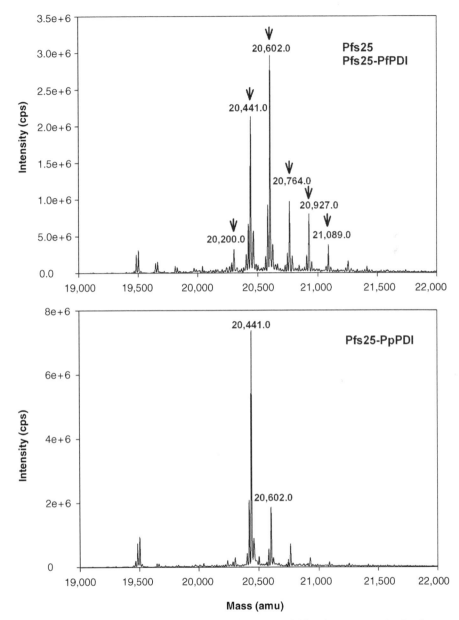

Figure 3.10. Analysis of purified Pfs25 (A) and Pfs25-PpPDI (A) by electron spray ionization mass spectrometry. Six peaks were identified as shown by the solid arrows. The initial peak contains no O-glycosylation (20,441 Da) and is followed by a series of five peaks containing 162 Da adducts of mannose. The IES-MS trace for Pfs25-PfPDI was similar to Pfs25 (not shown). Reprinted with permission from Elsevier [52].

TABLE 3.4. Select Critical Process Parameter and CQAs of Recombinant Malarial Drug Substances Manufactured at Pilot Scale for Phase I Human Clinical Trials

Critical Process or Quality Attribute	Outcome
Manufacturable process	Identified approach using *P. pastoris* for scalable production
Acceptable quality	Absence of N-linked glycosylation
	Limited O-linked mannosylation
	Structural integrity
	Induction of inhibitory antibodies
Acceptable quantity	>20 mg/L or ≥1000 doses/L broth
Overexpression of PDI	Improved profile for all of the above and reduction of O-linked mannosylation

ACKNOWLEDGMENTS

The work on expression and production of EBA-175 was performed by the Malaria Group at EntreMed, Inc., Rockville, MD. The Malaria Group was led by Dr. David Narum, Associate Director, and its members were Steve Fuhrmann, Steve Merchant, and Jennifer Wingard, along with Hong Liang from Molecular Biology and Dr. Tom Chen from Pharmaceutical Sciences. The Malaria Group was directed by Dr. B. Kim Lee Sim, Vice President, Preclinical Development. I also thank Dr. Louis Miller from the Malaria Vaccine Development Branch, DIR, NIAID, NIH, for providing the yeast-encoded synthetic gene sequence of AMA1-3D7. I appreciate the help of Drs Louis Miller, Matthew Plassmeyer, Laura Martin, and Mr. Richard Shimp for their critical review of the chapter. The EBA-175 project was funded in part with Federal funds from the National Institute of Allergy and Infectious Diseases, National Institutes of Health, under Contract No. AI-05421.

REFERENCES

[1] Snow RW, Guerra CA, Noor AM, Myint HY, Hay SI. The global distribution of clinical episodes of *Plasmodium falciparum* malaria. *Nature* 2005;434:214–217.

[2] Lengeler C. Insecticide-treated bed nets and curtains for preventing malaria. *Cochrane Database Syst Rev* 2004; CD000363.

[3] Moran M, Guzman J, Ropars AL, Jorgensen M, McDonald A, Potter S, Selassie HH. The Malaria Product Pipeline: Planning for the Future. The George Institute for International Health, 2007. http://www.thegeorgeinstitute.org/iih/media-and-publications/reports/2007.cfm.

[4] Aponte JJ, Aide P, Renom M, Mandomando I, Bassat Q, Sacarlal J, Manaca MN, Lafuente S, Barbosa A, Leach A, Lievens M, Vekemans J, Sigauque B, Dubois MC, Demoitie MA, Sillman M, Savarese B, McNeil JG, Macete E, Ballou WR, Cohen J, Alonso PL. Safety of the RTS,S/AS02D candidate malaria vaccine in infants living in a highly endemic area of

Mozambique: a double blind randomised controlled phase I/IIb trial. *Lancet* 2007;370: 1543–1551.

[5] Kocken CH, Withers-Martinez C, Dubbeld MA, van der Wel A, Hackett F, Valderrama A, Blackman MJ, Thomas AW. High-level expression of the malaria blood-stage vaccine candidate *Plasmodium falciparum* apical membrane antigen 1 and induction of antibodies that inhibit erythrocyte invasion. *Infect Immun* 2002;70:4471–4476.

[6] Miles AP, Zhang Y, Saul A, Stowers AW. Large-scale purification and characterization of malaria vaccine candidate antigen Pvs25H for use in clinical trials. *Protein Expr Purif* 2002;25:87–96.

[7] ICH. Draft consensus guideline, pharmaceutical development annex to Q8, current Step 2 version. International Conference on Harmonisation of Technical Requirements for Registration of Pharmaceuticals for Human Use, 2007.

[8] Yu LX. Pharmaceutical quality by design: product and process development, understanding, and control. *Pharm Res* 2008;25:781–791.

[9] Hwang WY, Foote J. Immunogenicity of engineered antibodies. *Methods* 2005;36:3–10.

[10] Shields RL, Lai J, Keck R, O'Connell LY, Hong K, Meng YG, Weikert SH, Presta LG. Lack of fucose on human IgG1 *N*-linked oligosaccharide improves binding to human Fcgamma RIII and antibody-dependent cellular toxicity. *J Biol Chem* 2002;277:26733–26740.

[11] Kubler-Kielb J, Majadly F, Wu Y, Narum DL, Guo C, Miller LH, Shiloach J, Robbins JB, Schneerson R. Long-lasting and transmission-blocking activity of antibodies to *Plasmodium falciparum* elicited in mice by protein conjugates of Pfs25. *Proc Natl Acad Sci USA* 2007;104:293–298.

[12] Tsai CW, Duggan PF, Shimp RL Jr, Miller LH, Narum DL. Overproduction of *Pichia pastoris* or *Plasmodium falciparum* protein disulfide isomerase affects expression, folding and O-linked glycosylation of a malaria vaccine candidate expressed in *P. pastoris*. *J Biotechnol* 2006;121:458–470.

[13] Darko CA, Angov E, Collins WE, Bergmann-Leitner ES, Girouard AS, Hitt SL, McBride JS, Diggs CL, Holder AA, Long CA, Barnwell JW, Lyon JA. The clinical-grade 42-kilodalton fragment of merozoite surface protein 1 of *Plasmodium falciparum* strain FVO expressed in *Escherichia coli* protects *Aotus nancymai* against challenge with homologous erythrocytic-stage parasites. *Infect Immun* 2005;73:287–297.

[14] Remarque EJ, Faber BW, Kocken CH, Thomas AW. Apical membrane antigen 1: a malaria vaccine candidate in review. *Trends Parasitol* 2008;24:74–84.

[15] Sim BK, Chitnis CE, Wasniowska K, Hadley TJ, Miller LH. Receptor and ligand domains for invasion of erythrocytes by *Plasmodium falciparum*. *Science* 1994;264:1941–1944.

[16] Narum DL, Haynes JD, Fuhrmann S, Moch K, Liang H, Hoffman SL, Sim BK. Antibodies against the *Plasmodium falciparum* receptor binding domain of EBA-175 block invasion pathways that do not involve sialic acids. *Infect Immun* 2000;68:1964–1966.

[17] Jones TR, Narum DL, Gozalo AS, Aguiar J, Fuhrmann SR, Liang H, Haynes JD, Moch JK, Lucas C, Luu T, Magill AJ, Hoffman SL, Sim BK. Protection of *Aotus* monkeys by *Plasmodium falciparum* EBA-175 region II DNA prime-protein boost immunization regimen. *J Infect Dis* 2001;183:303–312.

[18] Sim BK, Narum DL, Liang H, Fuhrmann SR, Obaldia N 3rd, Gramzinski R, Aguiar J, Haynes JD, Moch JK, Hoffman SL. Induction of biologically active antibodies in mice, rabbits, and monkeys by *Plasmodium falciparum* EBA-175 region II DNA vaccine. *Mol Med* 2001;7:247–254.

[19] Fraser TS, Kappe SH, Narum DL, VanBuskirk KM, Adams JH. Erythrocyte-binding activity of *Plasmodium yoelii* apical membrane antigen-1 expressed on the surface of transfected COS-7 cells. *Mol Biochem Parasitol* 2001;117:49–59.

[20] Kato K, Mayer DC, Singh S, Reid M, Miller LH. Domain III of *Plasmodium falciparum* apical membrane antigen 1 binds to the erythrocyte membrane protein Kx. *Proc Natl Acad Sci USA* 2005;102:5552–5557.

[21] Dicko A, Diemert DJ, Sagara I, Sogoba M, Niambele MB, Assadou MH, Guindo O, Kamate B, Baby M, Sissoko M, Malkin EM, Fay MP, Thera MA, Miura K, Dolo A, Diallo DA, Mullen GE, Long CA, Saul A, Doumbo O, Miller LH. Impact of a *Plasmodium falciparum* AMA1 vaccine on antibody responses in adult malians. *PLoS ONE* 2007;2:e1045.

[22] Malkin EM, Diemert DJ, McArthur JH, Perreault JR, Miles AP, Giersing BK, Mullen GE, Orcutt A, Muratova O, Awkal M, Zhou H, Wang J, Stowers A, Long CA, Mahanty S, Miller LH, Saul A, Durbin AP. Phase 1 clinical trial of apical membrane antigen 1: an asexual blood-stage vaccine for *Plasmodium falciparum* malaria. *Infect Immun* 2005;73:3677–3685.

[23] Wu Y, Ellis RD, Shaffer D, Fontes E, Malkin EM, Mahanty S, Fay MP, Narum D, Rausch KM, Miles AP, Aebig J, Orcutt A, Muratova O, Song G, Lambert L, Zhu D, Miura K, Long C, Saul A, Miller LH, Durbin AP. Phase 1 Trial of Malaria Transmission Blocking Vaccine Candidates Pfs25 and Pvs25 Formulated with Montanide ISA 51. *PLoS ONE* 2008;3 (8): e2940. http://www.plosone.org/doi/pone.0002636.

[24] Gardner MJ, Hall N, Fung E, White O, Berriman M, Hyman RW, Carlton JM, Pain A, Nelson KE, Bowman S, Paulsen IT, James K, Eisen JA, Rutherford K, Salzberg SL, Craig A, Kyes S, Chan MS, Nene V, Shallom SJ, Suh B, Peterson J, Angiuoli S, Pertea M, Allen J, Selengut J, Haft D, Mather MW, Vaidya AB, Martin DM, Fairlamb AH, Fraunholz MJ, Roos DS, Ralph SA, McFadden GI, Cummings LM, Subramanian GM, Mungall C, Venter JC, Carucci DJ, Hoffman SL, Newbold C, Davis RW, Fraser CM, Barrell B. Genome sequence of the human malaria parasite *Plasmodium falciparum*. *Nature* 2002;419:498–511.

[25] Liang H, Narum DL, Dey C, Chang A, Fuhrmann SR, Luu T, Sim BKL. Recombinant expression and characterization of EBA-175 receptor binding domain in *Pichia pastoris*. Program and Abstracts of the 48th Annual Meeting of the American Society of Tropical Medicine and Hygiene; 1999, Vol. 61. Abstract.

[26] Scorer CA, Buckholz RG, Clare JJ, Romanos MA. The intracellular production and secretion of HIV-1 envelope protein in the methylotrophic yeast. *Pichia pastoris*. *Gene* 1993;136:111–119.

[27] Kocken CH, van der Wel AM, Dubbeld MA, Narum DL, van de Rijke FM, van Gemert GJ, van der Linde X, Bannister LH, Janse C, Waters AP, Thomas AW. Precise timing of expression of a *Plasmodium falciparum*-derived transgene in *Plasmodium berghei* is a critical determinant of subsequent subcellular localization. *J Biol Chem* 1998;273:15119–15124.

[28] Ikemura T. Codon usage and tRNA content in unicellular and multicellular organisms. *Mol Biol Evol* 1985;2:13–34.

[29] Narum DL, Kumar S, Rogers WO, Fuhrmann SR, Liang H, Oakley M, Taye A, Sim BK, Hoffman SL. Codon optimization of gene fragments encoding *Plasmodium falciparum* merzoite proteins enhances DNA vaccine protein expression and immunogenicity in mice. *Infect Immun* 2001;69:7250–7253.

[30] Kennedy MC, Wang J, Zhang Y, Miles AP, Chitsaz F, Saul A, Long CA, Miller LH, Stowers AW. *In vitro* studies with recombinant *Plasmodium falciparum* apical membrane antigen 1 (AMA1): production and activity of an AMA1 vaccine and generation of a multiallelic response. *Infect Immun* 2002;70:6948–6960.

[31] Liang H, Narum DL, Fuhrmann SR, Luu T, Sim BK. A recombinant baculovirus-expressed *Plasmodium falciparum* receptor-binding domain of erythrocyte binding protein EBA-175 biologically mimics native protein. *Infect Immun* 2000;68:3564–3568.

[32] Thomas AW, Trape JF, Rogier C, Goncalves A, Rosario VE, Narum DL. High prevalence of natural antibodies against *Plasmodium falciparum* 83-kilodalton apical membrane antigen (PF83/AMA-1) as detected by capture-enzyme-linked immunosorbent assay using full-length baculovirus recombinant PF83/AMA-1. *Am J Trop Med Hyg* 1994;51:730–740.

[33] Narum DL, Liang H, Fuhrmann S, Sim BK. Synthetic Genes for Malarial Proteins and Methods of Use. United States Patent 7,078,507. 2006.

[34] Narum DL, Sim BK. Anti-plasmodium composition and methods of use. United States Patent 7,025,961. 2006.

[35] Gowda DC, Davidson EA. Protein glycosylation in the malaria parasite. *Parasitol Today* 1999;15:147–152.

[36] Stowers AW, Chen Lh LH, Zhang Y, Kennedy MC, Zou L, Lambert L, Rice TJ, Kaslow DC, Saul A, Long CA, Meade H, Miller LH. A recombinant vaccine expressed in the milk of transgenic mice protects *Aotus* monkeys from a lethal challenge with *Plasmodium falciparum*. *Proc Natl Acad Sci USA* 2002;99:339–344.

[37] Blackman MJ, Scott-Finnigan TJ, Shai S, Holder AA. Antibodies inhibit the protease-mediated processing of a malaria merozoite surface protein. *J Exp Med* 1994;180:389–393.

[38] Mayer DC, Kaneko O, Hudson-Taylor DE, Reid ME, Miller LH. Characterization of a *Plasmodium falciparum* erythrocyte-binding protein paralogous to EBA-175. *Proc Natl Acad Sci USA* 2001;98:5222–5227.

[39] Narum DL, Fuhrmann SR, Luu T, Sim BK. A novel *Plasmodium falciparum* erythrocyte binding protein-2 (EBP2/BAEBL) involved in erythrocyte receptor binding. *Mol Biochem Parasitol* 2002;119:159–168.

[40] Thompson JK, Triglia T, Reed MB, Cowman AF. A novel ligand from *Plasmodium falciparum* that binds to a sialic acid-containing receptor on the surface of human erythrocytes. *Mol Microbiol* 2001;41:47–58.

[41] Adams JH, Sim BK, Dolan SA, Fang X, Kaslow DC, Miller LH. A family of erythrocyte binding proteins of malaria parasites. *Proc Natl Acad Sci USA* 1992;89:7085–7089.

[42] Tolia NH, Enemark EJ, Sim BK, Joshua-Tor L. Structural basis for the EBA-175 erythrocyte invasion pathway of the malaria parasite *Plasmodium falciparum*. *Cell* 2005;122:183–193.

[43] Peek LJ, Brandau DT, Jones LS, Joshi SB, Middaugh CR. A systematic approach to stabilizing EBA-175 RII-NG for use as a malaria vaccine. *Vaccine* 2006;24:5839–5851.

[44] Avril M, Kulasekara BR, Gose SO, Rowe C, Dahlback M, Duffy PE, Fried M, Salanti A, Misher L, Narum DL Smith JD. Evidence for globally shared, cross-reacting polymorphic epitopes in the pregnancy-associated malaria vaccine candidate VAR2CSA. *Infect Immun* 2008;76:1791–1800.

[45] Sitia R, Braakman I. Quality control in the endoplasmic reticulum protein factory. *Nature* 2003;426:891–894.

[46] Noiva R. Protein disulfide isomerase: the multifunctional redox chaperone of the endoplasmic reticulum. *Semin Cell Dev Biol* 1999;10:481–493.

[47] Hsu TA, Watson S, Eiden JJ, Betenbaugh MJ. Rescue of immunoglobulins from insolubility is facilitated by PDI in the baculovirus expression system. *Protein Expr Purif* 1996;7:281–288.

[48] Kajino T, Ohto C, Muramatsu M, Obata S, Udaka S, Yamada Y, Takahashi H. A protein disulfide isomerase gene fusion expression system that increases the extracellular productivity of *Bacillus brevis*. *Appl Environ Microbiol* 2000;66:638–642.

[49] Kurokawa Y, Yanagi H, Yura T. Overproduction of bacterial protein disulfide isomerase (DsbC) and its modulator (DsbD) markedly enhances periplasmic production of human nerve growth factor in *Escherichia coli*. *J Biol Chem* 2001;276:14393–14399.

[50] Zhan X, Schwaller M, Gilbert HF, Georgiou G. Facilitating the formation of disulfide bonds in the *Escherichia coli* periplasm via coexpression of yeast protein disulfide isomerase. *Biotechnol Prog* 1999;15:1033–1038.

[51] Zou L, Miles AP, Wang J, Stowers AW. Expression of malaria transmission-blocking vaccine antigen Pfs25 in *Pichia pastoris* for use in human clinical trials. *Vaccine* 2003;21:1650–1657.

[52] Tsai CW, Duggan PF, Shimp RL Jr, Miller LH, Narum DL. Overproduction of *Pichia pastoris* or *Plasmodium falciparum* protein disulfide isomerase affects expression, folding and O-linked glycosylation of a malaria vaccine candidate expressed in *P. pastoris*. *J Biotechnol* 2006;121:458–470.

4

USING A RISK ASSESSMENT PROCESS TO DETERMINE CRITICALITY OF PRODUCT QUALITY ATTRIBUTES

Mark A. Schenerman, Milton J. Axley, Cynthia N. Oliver, Kripa Ram, and Gail F. Wasserman

4.1 INTRODUCTION

This chapter describes a process for determining the criticality of quality attributes for biotechnology-derived products. For these products, a phased implementation of the Quality by Design (QbD) principles [1–3] may be suitable because of increased complexity for the process and product compared to small drug molecules. As an example, a semiquantitative approach to risk assessment and criticality determination of product quality attributes is described here. The result of the criticality determination guides the implementation of a suitable testing plan and appropriate process controls. The result also facilitates development of a meaningful manufacturing process and operational targets. As more knowledge of the process and product is gained during clinical studies, a more comprehensive approach to QbD implementation, such as establishment of design space or use of PAT, becomes more feasible.

Examples of criticality determinations are also presented for two product quality attributes (glycosylation and deamidation) of anti-respiratory syncytial virus (RSV) monoclonal antibodies. These examples illustrate the type of data that might be considered to support determination of criticality for product quality attributes, principally prior product knowledge (laboratory, nonclinical, and clinical studies) and process capability.

Quality by Design for Biopharmaceuticals, Edited by A. S. Rathore and R. Mhatre
Copyright © 2009 John Wiley & Sons, Inc.

On the basis of the authors' experience and interpretation of the current regulatory guidances, some definitions are provided. Other relevant definitions can be found in ICH Q6B, Q8, and Q9.

> *Prior Product Knowledge:* the accumulated laboratory, nonclinical, and clinical experiences for a specific product quality attribute. This knowledge may also include relevant data from other similar molecules or from scientific literature.
>
> *Process Capability:* the demonstrated range of product attribute data on the basis of laboratory studies, manufacturing experience, and process control.
>
> *Noncritical Product Quality Attribute:* an attribute for which the combination of severity and probability results in a low risk of impact on safety or efficacy.
>
> *Key Product Quality Attribute:* an attribute for which the combination of severity and probability results in a moderate risk of impact on safety or efficacy. Typically, for this type of attribute, the range of prior product knowledge is broader than the process capability.
>
> *Critical Product Quality Attribute:* an attribute for which the combination of severity and probability results in a high or unknown risk of impact on safety or efficacy. There may be a potential for process capability to exceed the range of prior product knowledge, in the absence of adequate control.

4.1.1 Quality Attribute Risk Management

A risk assessment process is used to determine which quality attributes are important to the performance of the product (safety and efficacy) and to justify the acceptable range of variation for these attributes. The acceptable range or boundary for each quality attribute in turn provides the operational target designed to robustly control each attribute and assists in establishing the design space for manufacturing process parameters. All product quality attributes are considered and assessed for risk to determine the appropriate testing (e.g., routine monitoring, characterization testing, or no testing) required as part of the overall product-testing plan. A flowchart for the product quality attribute risk management is presented in Fig. 4.1 with the following outcomes.

- Product-related substances can be evaluated by using criticality determination and do not require routine monitoring.
- Product-related impurities are also evaluated by using criticality determination. However, these attributes often require routine monitoring.
- Risk assessment for process-related impurities is based on either criticality determination or safety assessment, and routine testing may not be necessary.
- Attributes already established as regulatory requirements are included in routine monitoring.

4.1.2 Criticality Determination

As part of the product quality risk assessment, criticality determination is used to identify the quality attributes that are important to the performance of the product (safety and

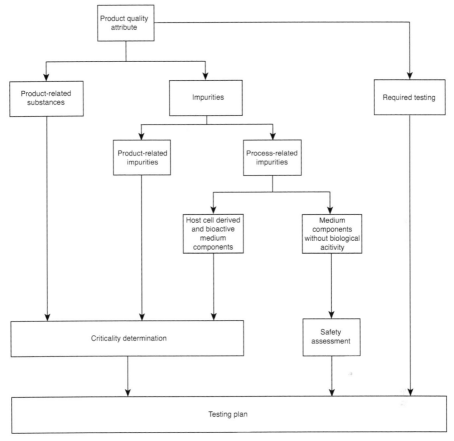

Figure 4.1. Product quality attribute risk management.

efficacy) and to evaluate the risk (severity and probability) for these attributes. This process follows the question-based guidance outlined in ICH Guidance for Industry Q9 Quality Risk Management:

- What might go wrong (attribute)?
- What are the consequences (severity)?
- What is the likelihood it will go wrong (probability)?

Figure 4.2 illustrates the outcome of the criticality determination for an attribute identified as potentially affecting product quality. Risk is defined in ICH Q9 as the product of the severity (consequences) and probability (likelihood of it going wrong). Thus, the assessment of severity and probability as shown in Fig. 4.2 provides the determination of criticality for an attribute.

Supporting information for this assessment includes prior product knowledge and process capability. Prior product knowledge is the accumulated laboratory, nonclinical, and clinical experiences for a specific product quality attribute. It may also include

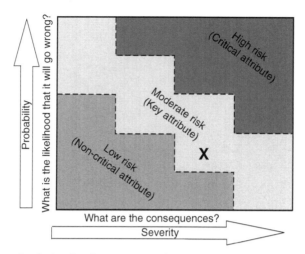

Figure 4.2. Example of criticality determination. (See the insert for color representation of this figure.)

relevant data from similar molecules and literature references. This combined knowledge provides rationale for relating the attribute to product safety and efficacy. Process capability is the demonstrated range of product attribute data on the basis of laboratory studies, manufacturing experience, and process control gathered during nonclinical and clinical development. Therefore, the capability of the process to deliver a product attribute within the range of prior product knowledge provides information on the probability of occurrence of harm. The risk assessment rationale together with data supporting the choice of the criticality of the attribute can be used to define a testing plan. The testing plan will be used to determine whether routine monitoring or characterization testing is conducted.

The first step in criticality determination gauges the severity of what might go wrong as the potential for the product quality attribute to impact safety (such as toxicity and immunogenicity) and efficacy (biological activity and pharmacokinetics (PK)). An "unknown" risk is considered equivalent to a "high" risk. The second step gauges the probability of an adverse impact, which is measured by the extent of the prior product knowledge compared to the process capability. The risk associated with the severity and the probability is a continuum. The method of assessment described here classifies the attributes as noncritical, key, and critical based on the following risk categories:

- *Low Risk:* In the case where the combination of severity and probability results in low risk, the attribute is considered "noncritical" and routine monitoring is not required. Testing for this attribute may be used to provide additional information on product characterization.
- *Moderate Risk:* In the case where the combination of severity and probability results in moderate risk, the attribute is considered "key." Depending on the extent of prior product knowledge (i.e., laboratory, nonclinical, and clinical data) compared to process capability, the attribute may be tested by (1) routine

monitoring with a specification that is equal to or broader than the process capability or (2) characterization only. If routine monitoring is performed and the attribute changes over time, the test may also be part of stability assessment.

- *High risk:* In the case where the combination of severity and probability results in high risk, the attribute is considered "critical" and a test for it will be a part of routine monitoring. The test may also be part of stability assessment if the attribute changes over time.

Semiquantitative plots of the criticality determination outcomes such as that shown in Fig. 4.2 can be used to score low, moderate, or high risk. The location of the "X" on each plot represents the combination of severity and probability, and this assessment is based on prior product knowledge, manufacturing experience, and scientific judgment. Nonclinical and clinical studies that show no adverse impact mitigate the severity and result in placement of the "X" toward the lower end of the severity scale. A high degree of process robustness and control or a process capability that is well within the range of prior product knowledge would mitigate the probability and result in placement of the "X" toward the lower end of the probability scale. The regions of the plot are colored green (low risk—noncritical attribute), yellow (moderate risk—key attribute), or red (high risk—critical attribute) to represent overall risk. The criticality of each attribute determines the testing plan.

4.1.3 Safety Assessment Strategy for Process-Related Impurities

Process-related impurities can be subdivided into two categories for the purposes of product quality attribute risk assessment, as shown in Fig. 4.1. Host cell-derived and bioactive components comprise a category of process-related impurities that undergo criticality determination in the same manner as performed for product-related substances and impurities. Examples of this category include host cell DNA, host cell protein, medium supplements such as protein hydrolysates from plant or microbial sources, or components such as nucleases or residual protein A from chromatography gels used in the process to facilitate purification.

Nonbioactive components such as antifoam are considered for their potential safety risk by evaluating an impurity safety factor (ISF). The ISF is the ratio of the impurity LD_{50} to the maximum amount of an impurity potentially present in the product dose:

$$ISF = LD_{50} \div \text{level in product dose}$$

where LD_{50} is the dose of an impurity that results in lethality in 50% of animals tested, and the level in product dose refers to the maximum amount of an impurity that could potentially be present and coadministered in a dose of product. Thus, ISF is a normalized measure of the relationship between the level of an impurity resulting in a quantifiable toxic effect and the potential exposure of a patient to impurity in the product. The higher the ISF, the greater the difference between the toxic effect and the potential product dose levels for an impurity, indicating therefore a minimized safety risk.

For the calculation of the ISF, the impurity level in a product dose is determined based on worst-case assumptions. In the absence of an assay to detect impurity, it is assumed that all of the impurity in the process copurifies with the product and no clearance is achieved by the purification process. Although this is a conservative assumption and unlikely to occur when orthogonal methods of separation are used in a purification process, it nevertheless allows calculation of the maximum potential content in the final product as a worst-case calculation. In the cases where a sufficiently sensitive assay is available, the actual level of an impurity in the product is determined on the basis of assay quantitation.

LD_{50} values can be found in the literature for many process-related impurities. Therefore, the LD_{50} values represent an available and quantitative indicator of acute toxicity that provides a useful comparator for assessing the risk posed by process-related impurity. However, the LD_{50} is a relatively imprecise measure of toxicity, and LD_{50} values are generally orders of magnitude higher than the levels of process-related impurities. Another measure of toxicity, the NOAEL (no observed adverse effects level), represents the level of a compound shown to be safe in animal experiments. NOAEL includes a longer term and more comprehensive assessment of organ system safety compared to acute lethality by LD_{50} measures. Because NOAEL is not readily available for most compounds, it cannot be routinely employed as a measure of safety. However, when both LD_{50} and NOAEL measures are available, they provide a link between safety and toxicity and are useful for the assessment of risk. Literature searches have revealed examples of compounds for which both NOAEL and the LD_{50} values are reported, and these examples show that NOAEL is generally one to two orders of magnitude below the LD_{50} values. On the basis of this rationale, we designated an ISF value of 1000 as representing a conservative estimate of safety where values at or above this threshold represent minimal risk.

The risk assessment strategy consists of a series of steps to evaluate impurity in terms of its risk to product safety. This process is outlined in Fig. 4.3 as a decision tree. Impurities can be eliminated from further consideration at any step where the safety risk is determined to be minimal.

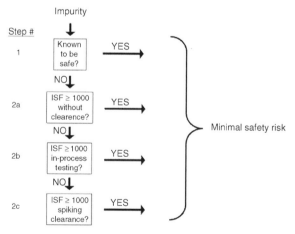

Figure 4.3. Representation of process-related impurities safety assessment strategy.

4.1.4 Testing Plan

The overall risk management plan, including the risk assessment (criticality determination and safety assessment) and the testing plan, is shown in Fig. 4.1. This testing plan reflects criticality determination, safety assessment, and regulatory requirements for product attributes. The attribute classification (high, medium, or low risk to product quality) serves as a rationale for the product testing plan (routine monitoring, characterization testing, no testing), specifications, and process controls that ensure minimal risk to product quality. The selection of tests appropriate for an attribute is based on the prior product knowledge and the overall control strategy. Each product quality attribute may be assessed by multiple tests. For example, monoclonal antibody fragments can be detected by SDS-PAGE, analytical ultracentrifugation, high-performance size-exclusion chromatography, and reversed-phase HPLC [4].

The specifications for a product can be established by using a combination of approaches, consistent with the guidance in ICH Q6B (Fig. 4.4). Wherever appropriate, certain attribute specifications are based on existing regulatory guidance. For other attributes, limits are established by considering both the prior product knowledge from

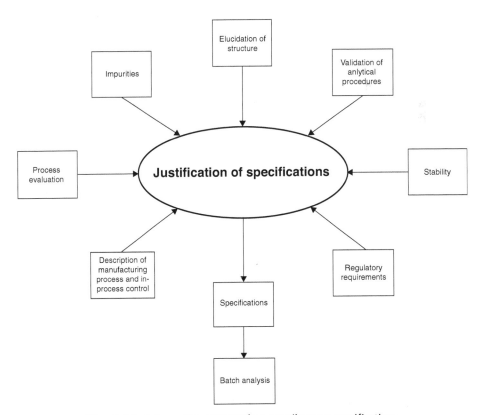

Figure 4.4. Information sources that contribute to specifications.

development studies and the process capabilities. In cases where there is limited prior product knowledge, specifications can be based on process capability data.

4.2 EXAMPLES OF CRITICALITY DETERMINATION

To illustrate the process by which the criticalities of product quality attributes can be determined, two examples are provided here: glycosylation and deamidation. These criticality determinations are described primarily for the anti-RSV monoclonal antibody motavizumab; however, relevant information for palivizumab is also included where appropriate. Both palivizumab and motavizumab target the F protein of RSV, and the molecules have been developed as prophylactics for prevention of RSV infection in the primary population of premature infants. Motavizumab is the "next generation" molecule developed through directed genetic engineering of the palivizumab molecule to achieve greater affinity for the target antigen [5]. Thus, motavizumab and palivizumab have the same mechanism of action, biological activity, and quality attributes, while differing by 13 amino acids. For these reasons, information for both molecules is combined to create a cumulative prior product knowledge and context for criticality determinations.

4.2.1 Case Study 1: Glycosylation

Prior Product Knowledge

LABORATORY STUDIES. Mammalian IgG1molecules generally have a conserved site of N-linked glycosylation in C_H2 domain of the Fc region of the heavy chains [6]. For motavizumab, Asn-300 on the heavy chain is the only consensus site for N-glycosylation and its occupancy was confirmed by peptide mapping with LC–MS.

The oligosaccharide structures of motavizumab are commonly found in monoclonal antibodies derived from NS0 mouse myeloma cell line [6–8]. An oligosaccharide profiling method [9, 10] was used to identify the labeled oligosaccharide structures by using MS or tandem MS. The predominant glycoforms (approximately 80%) are fucosylated biantennary complex-type oligosaccharides either with no terminal galactose residues (G0) or with monogalactosylated (G1) and digalactosylated forms (G2) (Figs. 4.5 and 4.6).

The minor complex-type glycoforms are truncated G0 and G1 without Fuc or GlcNAc, sialylated G1 and G2 containing N-glycolylneuraminic acid (NGNA), and G1 and G2 with murine Gal-α-1,3-Gal linkages (Fig. 4.5).

Most humans have circulating IgG antibodies directed against Gal-α-1,3-Gal moieties exposed to glycoproteins [11]. A recent report [12] proposes that the presence in patients of predose IgE directed against Gal-α-1,3-Gal may lead to anaphylaxis following administration of cetuximab. While this report raises safety concerns for any protein therapeutic that contains Gal-α-1,3-Gal linkages, we believe the risk is minimal for palivizumab and motavizumab because (1) palivizumab has a very low immunogenicity (less than 1% in over 4 million doses given), indicating that the small amount of Gal-α-1,3-Gal glycoforms present is not likely to induce an antigalactosyl immunogenic response; (2) Gal-α-1,3-Gal glycoforms of palivizumab and motavizumab are contained

Name	Structure	Mass	IUPAC Name (short form)
(A) Complex type			
G0-GlcNAc-Fuc		1234.2	GlcNAcβ-2Mana-3/6(Mana-6/3)Manβ-4GlcNAcβ-4GlcNAc
G0-GlcNAc		1380.3	GlcNAcβ-2Mana-3/6(Mana-6/3)Manβ-4GlcNAcβ-4(Fuca-6)GlcNAc
G0		1583.5	GlcNAcβ-2Mana-6(GlcNAcβ-2Mana-3)Manβ-4GlcNAcβ-4(Fuca-6)GlcNAc
G1-GlcNAc		1542.5	Galβ-4GlcNAcβ-2Mana-3/6(Mana-6/3)Manβ-4GlcNAcβ-4(Fuca-6)GlcNAc
G1		1745.7	Galβ-4GlcNAcβ-2Mana-3/6(GlcNAcβ-2Mana-6/3)Manβ-4GlcNAcβ-4(Fuca-6)GlcNAc
G1-GlcNAc+Gal		1704.6	Gala-3Galβ-4GlcNAcβ-2Mana-3/6(Mana-6/3)Manβ-4GlcNAcβ-4(Fuca-6)GlcNAc
G1+Gal		1907.8	Gala-3Galβ-4GlcNAcβ-2Mana-3/6(GlcNAcβ-2Mana-6/3)Manβ-4GlcNAcβ-4(Fuca-6)GlcNAc
G2		1907.8	Galβ-4GlcNAcβ-2Mana-6(Galβ-4GlcNAcβ-2Mana-3)Manβ-4GlcNAcβ-4(Fuca-6)GlcNAc
G1-GlcNAc+NGNA		1849.7	Neu5Gca2-3/6Galβ-4GlcNAcβ-2Mana-3/6(Mana-6/3)Manβ-4GlcNAcβ-4(Fuca-6)GlcNAc
G1+NGNA		2052.9	Neu5Gca2-3/6Galβ-4GlcNAcβ-2Mana-3/6(GlcNAcβ-2Mana-6/3)Manβ-4GlcNAcβ-4(Fuca-6)GlcNAc
G2+Gal		2069.9	Gala-3Galβ-4GlcNAcβ-2Mana-3/6(Galβ-4GlcNAcβ-2Mana-6/3)Manβ-4GlcNAcβ-4(Fuca-6)GlcNAc
G2+NGNA		2215.1	Neu5Gca2-3/6Galβ-4GlcNAcβ-2Mana-3/6(Galβ-4GlcNAcβ-2Mana-6/3)Manβ-4GlcNAcβ-4(Fuca-6)GlcNAc
G2+2Gal		2232.1	Gala-3Galβ-4GlcNAcβ-2Mana-3/6(Gala-3Galβ-4GlcNAcβ-2Mana-6/3)Manβ-4GlcNAcβ-4(Fuca-6)GlcNAc
G2+Gal+NGNA		2377.2	Neu5Gca2-3/6Galβ-4GlcNAcβ-2Mana-3/6(Gala-3Galβ-4GlcNAcβ-2Mana-6/3)Manβ-4GlcNAcβ-4(Fuca-6)GlcNAc
(B) High mannose type			
Man-5		1355.3	Mana-6(Mana-3)Mana-6(Mana-3)Manβ-4GlcNAcβ-4GlcNAc
Man-6		1517.4	Mana-6(Mana-3)Mana-6(Mana-3Mana-3)Manβ-4GlcNAcβ-4GlcNAc
(C) Hybrid type			
Hybrid 1704		1704.6	Mana-6(Mana-3)Mana-6(GlcNAcβ-2Mana-3)Manβ-4GlcNAcβ-4(Fuca-6)GlcNAc
Hybrid 1720		1720.6	Mana-6(Mana-3)Mana-6(Galβ-4GlcNAcβ-2Mana-3)Manβ-4GlcNAcβ-4GlcNAc
Hybrid 1866		1866.8	Mana-6(Mana-3)Mana-3/6(Galβ-4GlcNAcβ-2Mana-6/3)Manβ-4GlcNAcβ-4(Fuca-6)GlcNAc
Hybrid 2028		2028.9	Mana-3/6Mana-3/6(Mana-6/3)Mana-3/6(Galβ-4GlcNAcβ-2Mana-6/3)Manβ-4GlcNAcβ-4(Fuca-6)GlcNAc

● Man ■ GlcNAc ▲ Fuc ○ Gal ✦ 2-AB

Figure 4.5. Oligosaccharide structures. (See the insert for color representation of this figure.)

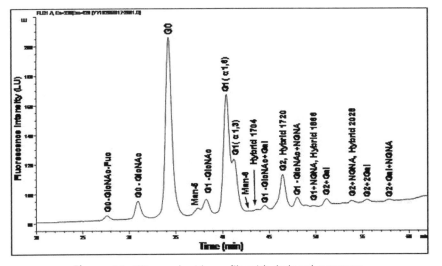

Figure 4.6. Oligosaccharide profile with deduced structures.

within the Fc portion of the molecule, while all the anti-Gal-α-1,3-Gal IgE binding to cetuximab was shown to be directed to the highly exposed sugars in the Fab variable region [12]; (3) typical N-linked oligosaccharides at Asn-297 of IgG antibodies, such as in palivizumab and motavizumab, are sequestered within the Fc domain polypeptide structure and are sterically hindered from interactions with other macromolecules [13]; (4) the presence of Gal-α-1,3-Gal glycoforms on the two Fab arms of the cetuximab molecule could provide a means by which cross-linking of IgE on the surface of mast cells could occur, resulting in a strong stimulus for the inflammatory and anaphylactic response [12]. The low levels of Gal-α-1,3-Gal glycoforms within the Fc domains of palivizumab and motavizumab would not be conducive to such cross-linking. (5) The intravenous route of administration of cetuximab may be responsible for the rapid hypersensitivity response; palivizumab and motavizumab are delivered by intramuscular injection.

Motavizumab and palivizumab have a small proportion of high mannose-type structures (Man-5 and Man-6) and negligible proportions of the hybrid-type oligosaccharides (Fig. 4.6). It is recognized that higher mannose forms (in glycoprotein populations) may have a faster rate of clearance because of binding to the mannose receptor [13]. However, the likelihood that the variations in these glycoforms in motavizumab or palivizumab could affect PK is low, because the levels of Man-5 and Man-6 are consistently low in manufactured lots.

In addition to MS, some peaks in the oligosaccharide profile were also identified by spiking with 2-AB-labeled G0, G2, and Man-5 standards. Two positional isomers of 2-AB-labeled G1 were identified through sequential exoglycosidase digestion steps by using α-N-acetylhexosaminidase and α-1-2,3 mannosidase [14], and the Gal-α-1,3-Gal linkages were confirmed by α-galactosidase digestion [15].

A number of glycoforms were generated by enzymatic remodeling. Treatment with α-galactosidase was used to remove terminal galactose sugars from the G2 and G1 forms and to generate an enriched G0 glycoform (74.4%). Treatment with α-galactosyltransferase and UDP galactose was used to extend both arms of the oligosaccharide chain and to generate an enriched G2 glycoform (69.7%). The oligosaccharide side chain was removed after treatment with PNGase F (deglycosylated motavizumab), and Fab and F(ab$'$)$_2$ were generated after treatment with pepsin and papain, respectively [16, 17] (Fig. 4.7).

Motavizumab variants were evaluated for F protein binding activity and RSV microneutralization (Table 4.1). The results demonstrated that the N-linked oligosaccharide side chain has no significant effect on binding activity to RSV F protein by ELISA and surface plasmon resonance (SPR) or microneutralization, and, furthermore, the Fc region is not required for RSV neutralizing activity.

These results are similar to those observed with palivizumab binding to F protein, as shown in Table 4.2. All the palivizumab variants had comparable F protein binding activity (by ELISA and SPR) as well as neutralizing activity, indicating that the oligosaccharides side chains do not affect the *in vitro* functional activity.

Similar studies carried out by Wu and coworkers [18] showed that both palivizumab and a palivizumab F(ab$'$)$_2$ fragment could inhibit RSV replication *in vitro* with inhibitory concentration (IC$_{50}$) values of 3.6 and 1.4 nM, respectively. This work further

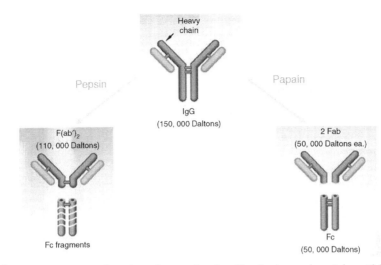

Figure 4.7. Protease digestion of monoclonal antibodies to produce Fab or F(ab')2.

TABLE 4.1. In Vitro Characterization of Motavizumab Variants

Sample Description	F Protein Binding ELISA (% of Reference Standard)	RSV Neutralization Activity (ED$_{50}$ Ratio)	SPR Binding (K_d) to F Protein (pM)
Control sample	98	1.0	4.0
Deglycosylated	102	1.0	6.2
G0 enriched	95	1.0	5.9
G2 enriched	112	0.9	5.0
IgG control	NA	No	NA
Fab	Binds to F protein[a]	Yes[a]	13.0
F(ab')$_2$	Binds to F protein[a]	Yes[a]	3.4

NA: not applicable.
[a]Because of expected differences in avidity for Fab and F(ab')$_2$ samples, it is not possible to directly compare them to reference standard at the same concentration.

TABLE 4.2. In Vitro Characterization of Palivizumab Variants

Sample Description	F Protein Binding ELISA (% of Reference Standard)	RSV Neutralization	SPR Binding (K_d) to F Protein (nM)
Control sample	98	Yes	0.81
Deglycosylated	98	Yes	0.88
G0 enriched	92	Yes	0.87
G2 enriched	98	Yes	1.04
Fab	Binds to F protein[a]	Yes	12.7
F(ab')2	Binds to F protein[a]	Yes	0.61

[a]Because of expected differences in avidity for Fab and F(ab')$_2$ samples, it is not possible to directly compare them to the reference standard at the same concentration.

TABLE 4.3. Binding Kinetics for Motavizumab Glycoforms to Human FcRn

Sample	SPR Binding (K_d) to Human FcRn (µM)
Motavizumab	2.26
G0 enriched	2.77
G2 enriched	2.18

demonstrates that the carbohydrate moiety of palivizumab is not required for virus neutralization activity.

The serum levels of IgG molecules *in vivo* are regulated primarily through binding of Fc portion of the antibody to FcRn receptor [19]. For this reason, we considered whether variation in glycoforms could affect FcRn binding to motavizumab. Binding kinetics for different glycoforms of motavizumab to human FcRn were analyzed by SPR, as shown in Table 4.3. The results show similar binding kinetics for motavizumab, G0-enriched, and G2-enriched motavizumab. Because the different glycoforms have similar binding to human FcRn, variation in glycosylation would not be expected to have an impact on PK.

NONCLINICAL STUDIES. The cotton rat has been the primary animal model for *in vivo* testing of RSV infection and disease [20, 21]. The cotton rat model has been used to demonstrate that passive immunization with anti-RSV antibodies can prevent RSV infection *in vivo*, and these results have supported the successful clinical testing and licensure of anti-RSV antibody prophylactics [22, 23]. For these reasons, the cotton rat model was used for testing the impact of glycoforms on motavizumab and palivizumab *in vivo* activities.

Motavizumab glycoforms and deglycosylated motavizumab were evaluated in a cotton rat PK study. Motavizumab levels were measured in serum (Fig. 4.8), bronchial alveolar lavage (BAL) (Fig. 4.9), and lung homogenate (Fig. 4.10). The results show that

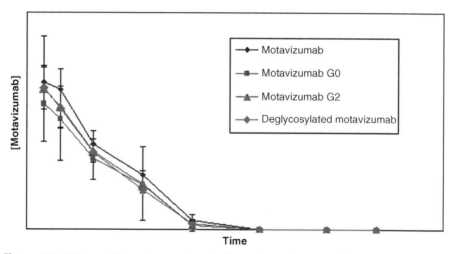

Figure 4.8. Cotton rat PK study: serum levels of motavizumab variants. (See the insert for color representation of this figure.)

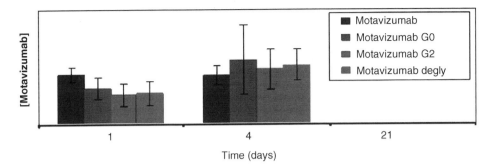

Degly = deglycosylated

Figure 4.9. Cotton rat PK study: BAL levels of motavizumab variants. Motavizumab is undetectable at 21 days. (See the insert for color representation of this figure.)

there was no statistical difference in the levels of motavizumab in serum, BAL, and lung homogenate for the variants tested.

Similar results were obtained previously with palivizumab variants. Palivizumab was administered to cotton rats and levels were measured in serum, BAL, and lung homogenates. The results showed no statistical difference in the levels of palivizumab in serum (Fig. 4.11), BAL, and lung homogenate (data not shown).

Published data have previously shown that antibody-mediated inhibition of RSV infection in a cotton rat model is not dependent on the Fc portion of the IgG1 molecule [24]. Similarly, we found that palivizumab Fab and $F(ab')_2$ variants protect against RSV infection in an *in vivo* cotton rat model (Fig. 4.12). Since the Fab and $F(ab')_2$ variants lack the Fc portion of the molecule and, therefore, are completely devoid of oligosaccharides, these data provide supportive evidence that oligosaccharides are not required for the mechanism of action of antibody-dependent neutralization of RSV activity.

Untreated motavizumab, G0-enriched, and G2-enriched samples were also evaluated as a prophylaxis for RSV infection in cotton rats. Rats were dosed with untreated,

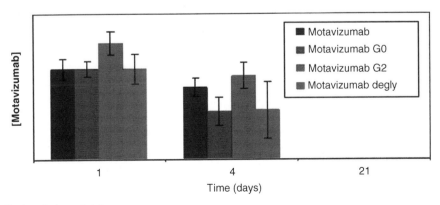

Degly = deglycosylated

Figure 4.10. Cotton rat PK study: lung homogenate levels of motavizumab variants. Motavizumab is undetectable at 21 days. (See the insert for color representation of this figure.)

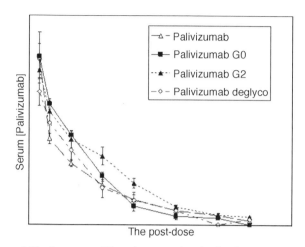

Figure 4.11. Cotton rat PK study: serum levels of palivizumab variants.

G0-enriched, or G2-enriched motavizumab, or bovine serum albumin (BSA) as a control. One group was challenged with RSV on Day 1 and the second on Day 3 postdosing with the respective protein. Four days later, all rats were sacrificed and the RSV titers in the lungs were measured. All motavizumab samples tested were found to provide >2.5 logs of RSV clearance when assessed at 5 and 7 days postdosing (Fig. 4.13). In addition, serum and lung levels of all forms were measured on the day the animals were sacrificed and the levels were found to be similar. These data demonstrate that changes in the structure of the oligosaccharide side chain do not affect potency and

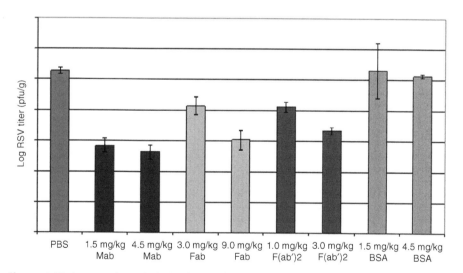

Figure 4.12. Intranasal prophylaxis of RSV infection in cotton rats by using palivizumab MAb, Fab, and F(ab')2 fragments. (See the insert for color representation of this figure.)

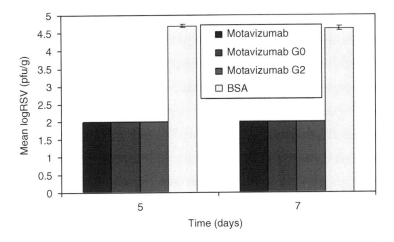

Lower limit of detection = 2 logs

Figure 4.13. IM prophylaxis in cotton rats by motavizumab-enriched glycoforms. (See the insert for color representation of this figure.)

PK in *in vivo* model. Glycoforms are effective in binding to the RSV, and the glycoforms remain in both the serum and lungs at concentrations necessary for viral neutralization.

Overall, the nonclinical *in vivo* studies show that oligosaccharides are not required for inhibition of RSV infection, and the variations in glycoforms do not affect pharmacokinetics and prophylaxis.

CLINICAL STUDIES. Variations in glycoform distribution may have a potential impact on the PK properties of the molecule. For example, an increased rate of clearance (shorter half-life) for a drug molecule results in decreased exposure of the drug to the target and therefore potentially a decrease in efficacy. Conversely, a decreased rate of clearance (longer half-life) increases exposure, which can increase the risk to safety if there are adverse effects of the drug. We evaluated whether variations in glycoform distribution could affect PK properties of motavizumab and palivizumab.

Analyses were carried out to assess the effect of glycoform distribution on the clinical PK parameter, area under the curve (AUC), as a measure of exposure. This evaluation was based on a statistical modeling approach for clinical serum data from palivizumab and motavizumab lots used in clinical studies. The data sets for palivizumab and motavizumab were considered together to determine the effect on PK because of the following reasons:

- *Similar Amino Acid Sequences:* There are 13 amino acid differences between motavizumab and palivizumab, and all are in the Fab region (Fig. 4.14) [5].
- *Same Glycoforms:* The structures of the glycoforms are the same but the proportions differ (Fig. 4.15).

	CDR-H1	CDR-H2	CDR-H3	FR-H4
Palivizumab	TSGMSVG	DIWWDDKKDYNPSLKS	SMITNWYFDV	WGAGTTVTVSS
Motavizumab	TAGMSVG	DIWWDDKKHYNPSLKD	DMIFNFYFDV	WGQGTTVTVSS

	CDR-L1	CDR-L2	CDR-L3	FR-L4
Palivizumab	KCQLSVGYMH	DTSKLAS	FQGSGYPFT	FGGGTKLEIK
Motavizumab	SASSRVGYMH	DTSKLAS	FQGSGYPFT	FGGGTKVEIK

Figure 4.14. Sequence differences between palivizumab and motavizumab.

In the oligosaccharide profile analysis (Dionex method) for palivizumab, the G1 and Man-5 glycoforms are not resolved. Therefore, analysis of palivizumab glycoforms included the three major observed peaks: G0, G2, and the combination G1/Man-5. Although the oligosaccharide profile (2-AB method) used for motavizumab glycoform analysis does resolve G1 and Man-5, for comparison to the palivizumab results the G1 and Man-5 values were combined for motavizumab. However, the 2-AB method has shown that Man-5 accounts for a low level (average of 7%) of the total G1 and Man-5 glycoforms, observed with the Dionex method, in palivizumab. The observed proportions of glycoforms for various lots of palivizumab and motavizumab are shown in Fig. 4.15.

Studies were performed to determine differences in clearance rates among the glycoforms of palivizumab and motavizumab. Product was isolated from human serum

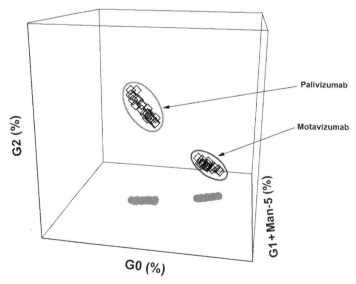

Figure 4.15. Observed glycoforms for motavizumab and palivizumab.

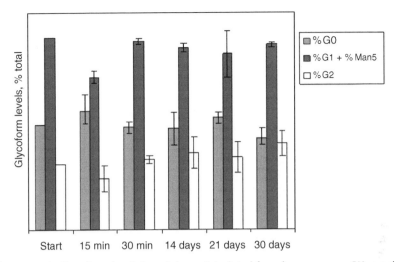

Figure 4.16. Glycoform levels in palivizumab isolated from human serum PK samples.

samples from clinical PK studies, and the major glycoform proportions in the isolated products were determined. An example of such results for palivizumab is shown in Fig. 4.16. We found no significant differences in the clearance rates for the major glycoforms within the constraints of the method.

A mathematical model was used to calculate the individual clearance rates of the major glycoforms in palivizumab and motavizumab, with a good fit of the data to the model. Antibody concentration in time course serum samples from PK studies and the relative oligosaccharide amount in the isolated antibody were used to develop this mathematical model. The estimated AUC of the glycoforms G0, G1 + Man-5, and G2 relative to drug product are shown in Fig. 4.17.

For the purpose of this model, the relative AUC estimates were computed, along with 90% confidence limits, at the outer bounds of the glycoform ranges that give rise to

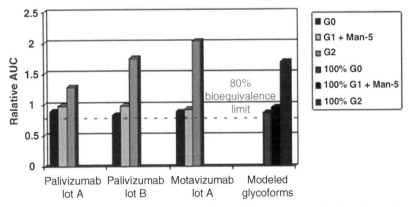

Figure 4.17. Modeling relative AUC for palivizumab and motavizumab glycoforms. (See the insert for color representation of this figure.)

motavizumab lots with AUCs no less than 80% of motavizumab drug product. The 80% lower limit is based on FDA guidance Statistical Approaches to Establishing Bioequivalence [25] to ensure no adverse impact on PK. The risk of an adverse impact on safety due to increased AUC is considered minimal for the following reasons: (1) nonclinical studies performed in cynomolgus monkeys found no safety signals at a dose 6.8 times higher than the human dose when given weekly for 26 weeks; (2) specificity of activity is directed against a nonhost target, that is, RSV; and (3) clinical trials conducted to date with motavizumab and palivizumab suggest a comparable safety profile. Therefore, an upper limit was deemed irrelevant.

The risk to efficacy is low based on the analysis because all the AUC values were ≥80% of the motavizumab values and therefore are considered bioequivalent exposure. The analysis supports the conclusion that any combination of oligosaccharide composition within the range of 0–100% for all three glycoforms (G0, G1 + Man-5, and G2) would result in a lot with an AUC no less than 80% of motavizumab drug product.

Palivizumab PK profiles were modeled for a typical multiple injection regimen for prophylactic administration covering the RSV season. The modeling was performed by using the derived half-life parameters for the palivizumab drug product and the hypothetical case of palivizumab consisting of 100% G2 glycoform. Figure 4.18 shows that the model predicts that 100% G2 form would display a PK profile that is well within the 95% confidence interval for palivizumab drug product on the basis of interpatient variability observed in clinical trials. Any difference in the PK of the extreme case of 100% G2 form is insignificant as compared to the natural variation in PK behavior among patients.

PROCESS CAPABILITY. The clinical lots that were tested had oligosaccharide profiles that were consistent with each other and the reference standard (Fig. 4.15). The results

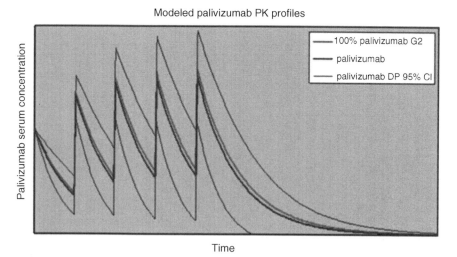

Figure 4.18. Seasonal modeling of palivizumab and G2 glycoform PK profiles. (See the insert for color representation of this figure.)

show that the oligosaccharide profile is consistent and reproducible at different scales (500 and 12,000 L) and different manufacturing sites.

CRITICALITY DETERMINATION. *Severity-Impact on Safety*: The major glycoforms shown in Fig. 4.5 are naturally occurring in humans and have no known safety risks. Furthermore, all glycoforms were present in all nonclinical and clinical studies, and an acceptable safety profile was observed.

Severity-Impact on Efficacy: Palivizumab and motavizumab are humanized monoclonal antibodies directed to an epitope in the A antigenic site of the F fusion protein of RSV. The sequences of motavizumab and palivizumab differ by 13 amino acids. The mechanism of action for motavizumab is the same as palivizumab. The Fab portion contains the viral binding elements, and the Fc portion contains an N-linked oligosaccharide chain attached to a single site on the heavy chain (Asn-300). The mechanism of action is through binding to the RSV F protein (specific neutralizing epitope) on the surface of the virion. The RSV F protein mediates an essential step of the infection process by enabling the fusion of the virus and host cell membranes, thus facilitating entry of the virus into the cell. Binding to the F protein blocks the essential site or sites for fusion to occur, thereby preventing infection of the cell by the virus.

The Fab-mediated mechanism of action for motavizumab is considerably simpler than that for other monoclonal antibodies (e.g., Herceptin® or Rituxan®) because these require Fc effector function-dependent activities, such as antibody-dependent cellular cytotoxicity or complement-dependent cytotoxicity [26–28]. Fc effector functions can be modulated by the glycosylation state [29] and consequently can affect the potency of such antibody products. In contrast, RSV neutralizing activity is not mediated by Fc effector functions, and changes in motavizumab Fc glycosylation patterns do not affect potency.

To evaluate the role of the N-linked oligosaccharide side chain on the biological properties of motavizumab and palivizumab, both antibodies were enzymatically remodeled to generate enriched G0, G2, and deglycosylated variants. All variants were evaluated in *in vitro* bioactivity assays as well as in the cotton rat animal model. Results for both antibodies support the conclusion that variations in the proportions of the major glycoforms found on the oligosaccharide side chain do not impact their biological properties. In addition, the results with palivizumab and motavizumab indicate that the PK profiles in cotton rats are not influenced by a range of oligosaccharide proportions. For human PK studies, any combination of oligosaccharide composition within the range of 0–100% for all three glycoforms (G0, G1 + Man-5, and G2) would result in a lot with AUC no less than 80% of motavizumab drug product. Taken together, these observations indicate that there is minimal risk to efficacy due to glycoforms proportions.

Likelihood: The manufacturing process consistently produces material that is well within the range of the prior product knowledge (Fig. 4.15). Therefore, the likelihood of occurrence is very low.

Conclusion: The criticality of glycosylation is shown in Fig. 4.19 and was determined by using the methodology previously described. On the basis of the data presented in this section, there is minimal risk to safety or efficacy due to variation in oligosaccharide proportions of motavizumab. Therefore, glycosylation is considered to be a

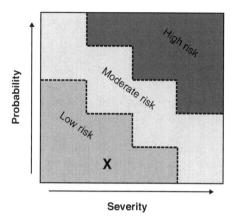

Figure 4.19. Criticality of glycosylation. (See the insert for color representation of this figure.)

noncritical quality attribute. The major oligosaccharide forms are defined as product-related substances, and therefore routine monitoring is not required.

4.2.2 Case Study 2: Deamidated Isoforms

Prior Product Knowledge

LABORATORY STUDIES. Deamidation at Asn or Gln residues is a common occurrence in human proteins [30, 31] and recombinant monoclonal antibodies [32]. Asn–Gly sequences are present and conserved in the constant regions of IgG molecules, and these sites are known to undergo deamidation under physiological conditions. The resulting charge isoform heterogeneity can be readily detected by using native IEF, which shows the formation of more acidic bands. The charge isoforms were further characterized by fractionating motavizumab using ion-exchange chromatography (IEC) as shown Fig. 4.20.

Peptide mapping of IEC fraction 1 (more acidic isoforms) and IEC fraction 2 with online mass spectrometry (MS) demonstrated that the major deamidation sites are located in the Fc region. The primary deamidation site is Asn-387 on the heavy chain as seen for other antibodies [33, 34]. Other identified deamidation sites in motavizumab that were detected at lower levels are Asn-157, Asn-289, Asn-318, Asn-328, Asn-424, and Asn-437. Because the bioactivity is independent of Fc function, deamidation is unlikely to have an effect on the activity of the molecule. As expected, the deamidated isoforms (IEC fraction 1) exhibited similar F protein binding activity to motavizumab and IEC fraction 2, as measured by ELISA and SPR, and were able to neutralize RSV (Table 4.4).

Motavizumab was also incubated in human plasma at 37°C for up to 5 weeks and analyzed by both IEC and native IEF. Native IEF analysis demonstrated that acidic bands were generated during incubation in human plasma at 37°C (Fig. 4.21). IEC analysis also showed an increase in acidic isoforms with time.

To demonstrate that the acidic isoforms observed by IEC and native IEF were due to deamidation, peptide mapping was performed on all samples recovered from human

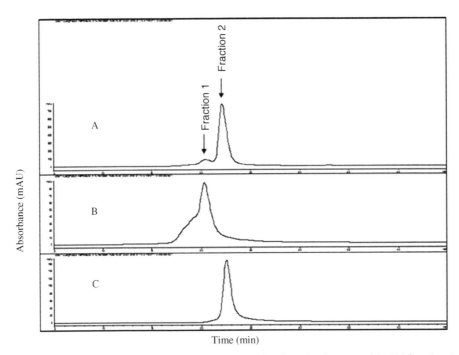

Figure 4.20. Fractionation by IEC. Panel A: IEC profile of motivation; panel B: IEC fraction 1; panel C: IEC fraction 2.

TABLE 4.4. Biological Characterization and IEC Fractions

Sample	F Protein Binding ELISA (%)[a]	RSV Microneutralization	F Protein Binding Kinetics (SPR) K_d (pM)
Motavizumab (unfractionated)	114	Neutralizes RSV	3.87
IEC fraction 1	112	Neutralizes RSV	4.58
IEC fraction 2	104	Neutralizes RSV	4.17

[a]Value listed is % binding relative to the reference standard.

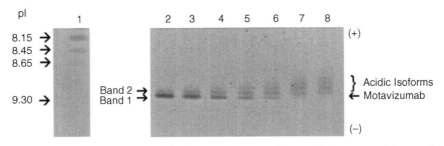

Figure 4.21. Native IEF analysis of motavizumab spiked in human plasma. Lanes: (1) pI marker; (2) reference standard; (3) initial time point; (4) 1 week in human plasma; (5) 2 weeks in human plasma; (6) 3 weeks in human plasma; (7) 4 weeks in human plasma; (8) 5 weeks in human plasma.

plasma after incubation at 37°C. The results confirmed that the acidic isoforms were due to deamidation and identified the primary site of deamidation as Asn-387, which is located on the H37 peptide of the heavy chain in the Fc region (Table 4.5). The samples were also characterized by densitometry of the native IEF gels and by analysis of binding and biological activity. Native IEF was performed on the same samples and the ratios of band 1 and band 2 ODs were determined by densitometry. The "OD ratios" from native IEF gels represent the level of deamidation as the ratio of optical density values determined by densitometry for the major product band (represented by band 1) and acidic isoform bands (represented by band 2). Lower OD ratio values indicate increased proportions of the acidic isoforms and, therefore, increased deamidation levels and vice versa. Total deamidation of motavizumab ranged from 25.4% at the initial time point to 77.2% after 5 weeks of exposure to human plasma at 37°C.

Deamidated motavizumab (up to 77.2% deamidation) exhibited F protein binding activity, as measured by ELISA and SPR, and was able to neutralize RSV. For these assays, the variability was greater than typically observed in lot release, characterization, and stability testing because the antibody isolated from the human serum was at very low concentrations. For SPR, the k_{off} was evaluated because it was the best indicator of a change in the ability to bind to either F protein because the SPR off-rate is independent of protein concentration. All k_{off} values were comparable to each other and to the initial value (Table 4.5). To overcome the limitations of the low protein concentration in the plasma incubation study, a control study was conducted by incubating higher protein concentration motavizumab samples at 37°C in pH 8.5 buffer for a period of up to 5 weeks (Table 4.6). Deamidation sites were confirmed to be the same as described earlier for the human plasma incubation site. In this study, binding activity was observed for deamidation levels of up to 79.0%. Hence, the buffer-incubated deamidated preparation was subsequently used in the nonclinical PK and bioactivity studies described later.

In the plasma incubation study, OD ratios as low as 0.25 (with a pI of greater than 8.65) resulted in some bioactivity. In the buffer incubation study, a similar level of deamidation (as determined by peptide mapping) resulted in an OD ratio of 0.50 (with a pI of greater than 8.65) and no reduction in binding to F protein. Because there is variability in the OD ratio at high deamidation levels, an OD ratio of 0.50 is presently designated as the lower OD ratio limit of the prior product knowledge for an acceptable level of deamidation.

The results of the incubation studies support the conclusion that the deamidation occurs naturally in human plasma. For OD ratios greater than 0.5 and pI values greater than 8.65, deamidation does not impact motavizumab binding or biological activity.

Nonclinical and Clinical Studies—Motavizumab. The level of deamidation in motavizumab produced under typical manufacturing conditions is approximately 20%, and it increases upon incubation in plasma. Motavizumab deamidated by incubation in buffer at pH 8.5 was evaluated in a cotton rat PK study. Deamidated motavizumab (1- and 5-week incubation), as well as unmodified motavizumab, was administered to groups of rats. Motavizumab levels were measured in serum (Fig. 4.22), BAL (Fig. 4.23), and lung homogenates (Fig. 4.24). The results showed no differences in the levels of motavizumab compared to deamidated motavizumab in serum, BAL, and lung.

TABLE 4.5. Identified Sites of Deamidation and Characterization after Incubation in Human Plasma

| | Identified Sites of Deamidation | | | | | Characterization | | | |
Samples	Asn-289 (%)	Asn-318 (%)	Asn-387 (%)	Total Deamidation (%)	Native IEF Band 1/Band 2 OD Ratio	F protein Binding ELISA (%)[a]	RSV Microneutralization	F protein Binding Kinetics (SPR) (k_{off}) (s^{-1})[b]
Initial	0	3.9	14.0	17.9	3.4	47	Neutralizes RSV	4.96×10^{-6}
1 week	1.6	4.6	19.2	25.4	2.1	88	Neutralizes RSV	6.36×10^{-6}
1 week	0	4.5	24.9	29.4	1.6	85	Neutralizes RSV	5.57×10^{-6}
2 weeks	2.8	4.0	34.5	41.3	1.2	91	Neutralizes RSV	6.07×10^{-6}
3 weeks	2.0	4.3	40.1	46.4	0.75	52	Neutralizes RSV	5.87×10^{-6}
4 weeks	4.8	3.6	47.2	55.6	0.40	77	Neutralizes RSV	5.81×10^{-6}
5 weeks	12.1	4.8	60.3	77.2	0.25			

Note: 0, 3.9, 14.0, and 17.9 are reference standards.

[a]Value listed is % binding relative to the reference standard.

[b]Samples extracted from human plasma had low total protein concentrations. Because the SPR off-rate is independent of protein concentration, it was used to analyze these samples.

TABLE 4.6. Characterization After Deamidation in pH 8.5 Buffer

| | Identified Sites of Deamidation | | | | | Characterization | | |
Samples	Asn-289 (%)	Asn-318 (%)	Asn-387 (%)	Total Deamidation (%)	Native IEF Band 1/Band 2 OD Ratio	F Protein Binding ELISA (%)[a]	RSV Microneutralization	F Protein Binding Kinetics (SPR) (k_{off}) (s^{-1})[b]
Initial	0	3.9	14.0	17.9	3.4	105	Neutralizes RSV	3.71×10^{-6}
1 week	1.4	4.9	12.3	18.6	2.3	103	Neutralizes RSV	4.66×10^{-6}
2 weeks	2.1	4.8	27.1	34.0	1.4	107	Neutralizes RSV	4.31×10^{-6}
3 weeks	2.1	6.5	35.5	44.1	1	90	Neutralizes RSV	4.55×10^{-6}
4 weeks	2.7	5.7	46.6	55.0	0.71	101	Neutralizes RSV	4.55×10^{-6}
5 weeks	3.8	4.8	53.2	61.8	0.71	100	Neutralizes RSV	4.93×10^{-6}
	3.4	5.8	55.9	65.1	0.5			

Note: 0, 3.9, 14.0, and 17.9 are reference standards.

[a]Value listed is % binding relative to reference standard.

[b]Samples extracted from human plasma had low total protein concentrations. Because the SPR off-rate is independent of protein concentration, it was used to analyze these samples.

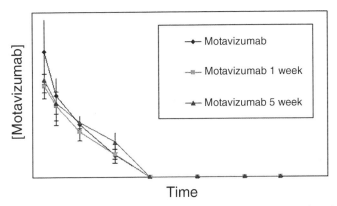

Figure 4.22. Cotton rat PK study: levels of motavizumab in serum after dosing.

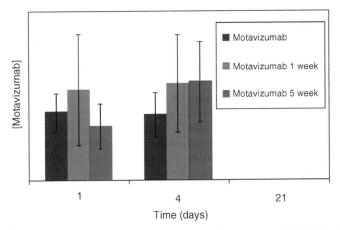

Figure 4.23. Cotton rat PK study measuring levels of motavizumab in BAL after dosing. Motavizumab is undetectable at 21 days.

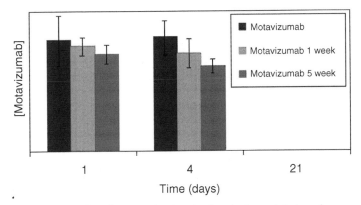

Figure 4.24. Cotton rat PK study measuring levels of motavizumab in lung homogenate after dosing. Motavizumab is undetectable at 21 days.

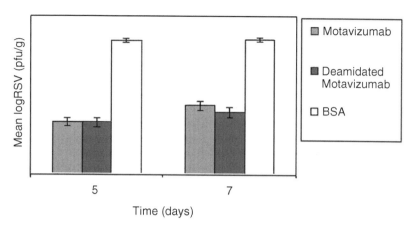

Figure 4.25. IM prophylaxis in cotton rats by deamidated motavizumab.

Deamidated motavizumab was also evaluated as a prophylaxis for RSV infection in cotton rats. Rats were dosed with deamidated motavizumab (4 weeks at pH 8.5), motavizumab, or BSA as a control. One group was challenged with RSV on Day 1 and the second on Day 3 postdosing with the respective protein. Four days post challenging, all rats were sacrificed and lung RSV titers were measured. Both deamidated and nondeamidated motavizumab provided 2.5 logs of RSV clearance when assessed at 5 and 7 days postdosing (Fig. 4.25). Serum and lung levels of motavizumab and deamidated motavizumab were measured on the day the animals were sacrificed and levels were found to be similar. These data indicate that deamidated motavizumab remains potent in an *in vivo* model, is effective in binding to the RSV, and the antibody remains in both serum and lungs at concentrations necessary for viral neutralization.

Although deamidation in the complementarity determining regions (CDRs) may affect antigen binding, no deamidation sites are present in the motavizumab CDRs. There are no known literature reports of immunogenicity in monoclonal antibodies linked to deamidation. The impact of deamidation on immunogenicity was evaluated in nonclinical and clinical studies. Cynomolgus monkeys were administered doses sixfold higher than the clinical dose at weekly intervals for 6 months and were thus exposed to far higher levels of deamidated product than would occur with the typical five monthly clinical doses. No immunogenicity or adverse reactions were observed.

NONCLINICAL AND CLINICAL STUDIES—PALIVIZUMAB. Similar to the results for motavizumab, palivizumab purified from clinical PK samples showed an increase in acidic isoforms indicative of deamidation (Fig. 4.26). Extensive studies were performed to verify that deamidated palivizumab retains biological activity. These studies showed that both normal and buffer-deamidated palivizumab retain biological activity as determined by the primary potency assay (F protein binding ELISA, 109 and 113% relative potency, respectively), as well as the relevant *in vitro* (RSV microneutralization, Fig. 4.27) and *in vivo* (cotton rat PK and cotton rat challenge) bioassays (Figs 4.28 and 4.29).

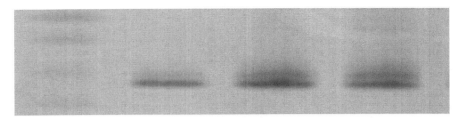

<u>Figure 4.26.</u> Native IEF analysis of palivizumab isolated from human PK samples. Lanes: (1) pI marker; (2) palivizumab reference standard; (3) palivizumab 15 min PK sample; (4) palivizumab 7 days PK sample.

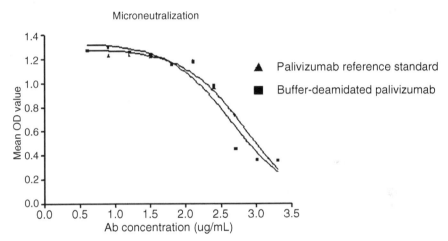

<u>Figure 4.27.</u> Comparison of palivizumab reference standard and buffer-deamidated material in the RSV microneutralization assay.

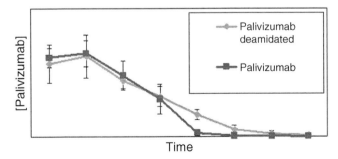

<u>Figure 4.28.</u> Cotton rat PK analysis of serum palivizumab levels after dosing with untreated and buffer-deamidated palivizumab.

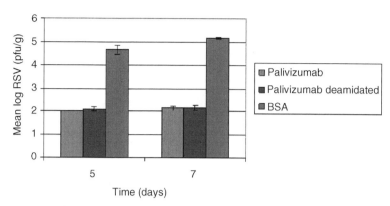

Figure 4.29. Prophylaxis of RSV infection in cotton rats by using palivizumab and buffer-deamidated palivizumab. (See the insert for color representation of this figure.)

Results of nonclinical and clinical studies support that deamidation of motavizumab and palivizumab is a minimal risk because (1) deamidation occurs naturally under physiological conditions following dosing of motavizumab or palivizumab to patients and therefore the resulting charge isoforms were evaluated during clinical safety and efficacy trials; (2) deamidation occurs at common sites of the Fc regions in motavizumab and palivizumab, which are not involved in the mechanism of action of the product; and (3) deamidation does not impact the potency or pharmacokinetics of motavizumab or palivizumab.

PROCESS CAPABILITY. Multiple lots were analyzed by native IEF and showed band patterns consistent with reference standard. On the basis of process experience, the Band 1/ Band 2 OD ratio values for all lots is greater than or equal to 1.4.

CRITICALITY DETERMINATION. *Severity-Impact on Safety*: On the basis of the fact that motavizumab is normally deamidated *in vivo*, the safety risk from immunogenicity as a result of deamidation is minimal.

Severity-Impact on Efficacy: It has been demonstrated that deamidated forms at OD ratios > 0.50 and with a pI greater than 8.65 exhibit biological activity (cotton rat challenge assay and RSV microneutralization), potency (F protein binding ELISA), and ligand binding capability to F protein (SPR). The risk to efficacy is minimal within this range of prior product knowledge.

Likelihood: The process capability (OD ≥ 1.4) is within the range of the prior product knowledge (OD ≥ 0.50). Therefore, there is a moderate likelihood of occurrence.

Conclusion: Based on the risk assessment, the criticality was determined to be moderate (Fig. 4.30), resulting in the classification of deamidated isoforms as a key quality attribute with routine monitoring in the testing plan.

Native IEF can be used to assess the identity and purity of the product. In additional, because the product is known to deamidate under accelerated temperature conditions, this assay may also be a stability monitoring tool. Hence, native IEF

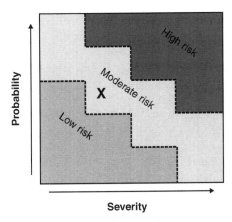

Figure 4.30. Criticality of deamidation. (See the insert for color representation of this figure.)

would be recommended for lot release and stability testing. By using this, the range of prior product knowledge is defined by the lower boundary of the development experience (0.50 OD ratio). The upper OD ratio boundary is not relevant because higher OD ratios indicate lower deamidation levels. Because the prior product knowledge (0.50 OD ratio) is broader than the process capability (>1.4 OD ratio), the data support a native IEF OD ratio specification less than 1.4 and greater than 0.50. The native IEF sample bands must remain within the pI range of prior product knowledge (pI > 8.65).

4.3 CONCLUSION

We describe here a risk assessment process to evaluate product quality attributes and categorize the attributes as being of low, medium, or high risk to product quality. The categorization of risk and the information obtained during the risk assessment process for an attribute provide the basis for determining the appropriate testing plan, including the test(s) performed, frequency of testing, and specifications.

In this chapter, we have illustrated the use of a risk assessment process with two examples of the glycosylation and deamidation quality attributes. The information was obtained during the development of two anti-RSV monoclonal antibodies (palivizumab and motavizumab). We show here that glycosylation is a noncritical (low-risk) attribute and deamidation is a key (moderate-risk) attribute. It should be noted that the assessments presented here are limited to these monoclonal antibodies. For other monoclonal antibodies, the outcome of the assessment might be different.

A risk assessment process for evaluating product quality attributes requires an accumulation of considerable information to achieve a meaningful outcome. The criticality determination necessitates a comparison of the prior product knowledge with the process capability. Some relevant information for prior product knowledge might be sourced from literature, but in many cases the data comes from nonclinical and clinical studies. Often this information is limited by the number of nonclinical and clinical studies performed, as well as by the range of variation for an attribute in the batches evaluated. As described here, in some cases prior product knowledge can be expanded beyond these limitations by inducing an intentional increase in attribute variant levels, followed by appropriate testing. Similarly, process capability information is limited to the number of product batches manufactured by the relevant process and tested for the attribute. Supportive information can be gleaned from similar processes for similar products. Platform production processes, in which the same process is used for production of multiple, similar products, can provide substantial amounts of predictive information for process capability.

The purpose of this risk management process is to identify the most appropriate testing plan for each product quality attribute. Risk assessment provides a rationale for the testing plan that enables efficient use of available resources, while assuring product quality.

ACKNOWLEDGMENTS

The authors gratefully acknowledge the outstanding technical and editorial contributions made by Orit Scharf, Patrick McGeehan, Gerard Lacourciere, David Robbins, Yuan Chang, Ziping Wei, Robert Strouse, Patricia Cash, Jose Casas-Finet, and Harry Yang. We would also like to thank Helen Jeanes, Nancy Craighead, and Anita Ymbert for their expert assistance in preparing the manuscript.

REFERENCES

[1] ICH Q6B. ICH Harmonised Tripartite Guideline on Specifications: Test Procedures and Acceptance Criteria for Biotechnological/Biological Products, 1999. Available at http://www.ich.org/LOB/media/MEDIA432.pdf

[2] ICH Q8. ICH Harmonised Tripartite Guideline on Pharmaceutical Development, 2005. Available at http://www.ich.org/LOB/media/MEDIA1707.pdf.

[3] ICH Q9. ICH Harmonised Tripartite Guideline on Quality Risk Management, 2005. Available at http://www.ich.org/LOB/media/MEDIA1957.pdf.

[4] Cordoba AJ, Shyon BJ, Breen D, Harris RJ. Non-enzymatic hinge region fragmentation of antibodies in solution. *J Chromatogr B Anal Technol Biomed Life Sci* 2005;818:115–121.

[5] Wu H, Pfarr DS, Johnson S, Brewah YA, Woods RM, Patel NK, White WI, Young JF, Kiener PA. Development of motavizumab, an ultra-potent antibody for the prevention of respiratory syncytial virus infection in the upper and lower respiratory tract. *J Mol Biol* 2007; 368:652–665.

[6] Jefferis R. Glycosylation of human IgG antibodies. *BioPharm Int* 2001;14:19–27.

[7] Yu Ip CC, Miller WJ, Silberklang M, Mark GE, Ellis RW, Huang L, Gluska J, Van Halbeek H, Zhu J, Alhadeff JA. Structural characterization of N-glycans of a humanized anti-CD18 murine immunoglobulin G. *Arch Biochem Biophys* 1994;308:387–399.

[8] Roberts GD, Johnson WP, Burman S, Aumula KR, Cart SA. An integrated strategy for structural characterization of the protein and carbohydrate components on monoclonal

antibodies: application to anti-respiratory syncytial virus Mab. *Anal Chem* 1995;67:3613–3625.

[9] Bigge JC, Patel TP, Bruce JA, Goulding PN, Charles SM, Parekh RB. Nonselective and efficient fluorescent labeling of glycans using 2-amino benzamide and anthranilic acid. *Anal Biochem* 1995;230:229–238.

[10] Guile GR, Rudd PM, Wing DR, Prime SB, Dwek RA. A rapid high-resolution high-performance liquid chromatographic method for separating glycan mixtures and analyzing oligosaccharide profiles. *Anal Biochem* 1996;240:210–226.

[11] Galili U. The alpha-gal epitope and the anti-Gal antibody in xenotransplantation and in cancer immunotherapy. *Immunol Cell Biol* 2005;83:674–686.

[12] Chung CH, Mirakhur B, Chan E, Le QT, Berlin J, Morse M, Murphy BA, Satinover SM, Hosen J, Mauro D, Slebos RJ, Zhour Q, Gold D, Hatley T, Hicklin DJ, Platts-Mills TA. Cetuximab-induced anaphylaxis and IgE specific for galactose-alpha-1,3-galactose. *N Engl J Med* 2008;358:1109–1117.

[13] Jones AJS, Papac DI, Chin EH, Keck R, Baughman SA, Lin YS, Kneer J, Battersby JE. Selective clearance of glycoforms of a complex glycoprotein pharmaceutical caused by terminal N-acetylglucosamine is similar in humans and cynomolgus monkeys. *Glycobiology* 2007;17:529–540.

[14] Prime S, Merry T. Exoglycosidase sequencing of N-linked glycans by the reagent array analysis methods (RAAM). In: Hounsell EF, editor. Methods in Molecular, Biology: Glycoanalysis Protocols. Humana Press; 1998. p 53–69.

[15] Spiro RG, Study of the carbohydrates of glycoproteins. *Methods Enzymol* 1972;28:3–43.

[16] Mage MG, Preparation of Fab fragments from IgGs of different animal species. *Methods Enzymol* 1980; 70: 142–150.

[17] Rousseaux J, Rousseaux-Prevost R, Bazin H. Optimal conditions for the preparation of proteolytic fragments from monoclonal IgG of different rat IgG subclasses. *Methods Enzymol* 1986;121:663–669.

[18] Wu H, Pfarr DS, Tang Y, An L-L, Patel NK, Watkins JD, Huse WD, Kiener PA, Young JF. Ultra-potent antibodies against respiratory syncytial virus: effects of binding kinetics and binding valence on viral neutralization. *J Mol Biol* 2005;350:126–144.

[19] Ghetie V, Ward ES. Multiple roles for the major histocompatibility complex class I-related receptor FcRn. *Annu Rev Immunol* 2000;18:739–766.

[20] Prince GA, Jenson AB, Horswood RL, Camargo E, Chanock RM. The pathogenesis of respiratory syncytial virus infection in cotton rats. *Am J Pathol* 1978;93:771–791.

[21] Byrd LG, Prince GA. Animal models of respiratory syncytial virus infection. *Clin Infect Dis* 1997;25:1363–1368.

[22] PREVENT Study Group. Reduction of respiratory syncytial virus hospitalization among premature infants and infants with bronchopulmonary dysplasia using respiratory syncytial virus immune globulin prophylaxis. *Pediatrics* 1997;99:93–99.

[23] IMPACT-RSV Study Group. Palivizumab, a humanized respiratory syncytial virus monoclonal antibody, reduces hospitalization from respiratory syncytial virus infection in high-risk infants. *Pediatrics* 1998;102:531–537.

[24] Prince GA, Hemming VG, Horswood RL, Baron PA, Murphy BR, Chanock RM. Mechanism of antibody-mediated viral clearance in immunotherapy of respiratory syncytial virus infection of cotton rats. *J Virol* 1990;64:3091–3092.

[25] FDA. Guidance for Industry on Statistical Approaches to Establishing Bioequivalence, 2001. Available at http://www.fda.gov/Cder/guidance/3616fnl.htm.

[26] Cartron G, Watier H, Golay J, Solal-Celigny P. From the bench to the bedside: ways to improve rituximab efficacy. *Blood* 2004;104:2635–2642.

[27] Nahta R, Esteva FJ. Herceptin: mechanisms of action and resistance. *Cancer Lett* 2006; 232:123–138.

[28] Musolino A, Naldi N, Bortesi B, Pezzuolo D, Capelletti M, Missale G, Laccabue D, Zerbini A, Camisa R, Bisagni G, Neri TM, Ardizzoni A. Immunoglobulin G fragment C receptor polymorphisms and clinical efficacy of Trastuzumab-based therapy in patients with HER-2/new-positive metastatic breast cancer. *J Clin Oncol* 2008;26:1789–1796.

[29] Jefferis R. Glycosylation of recombinant antibody therapeutics. *Biotechnol Prog* 2005;21:11–16.

[30] Huang LH, Li JR, Wroblewski VJ, Beals JM, Riggin RM. *In vivo* deamidation characterization of monoclonal antibody by LC/MS/MS. *Anal Chem* 77:2005;1432–1439.

[31] Lindner H, Helliger W. Age-dependent deamidation of asparagine residues in protein. *Exp Gerontol* 2001;36 (9):1551–1563.

[32] Tsai PK, Bruner MW, Irwin JI, Yu Ip CC, Oliver CN, Nelson RW, Volkin DB, Middaugh CR. Origin of the isoelectric heterogeneity of Monoclonal Immunoglobulin h1B4. *Pharm Res* 1993;10:1480–1586.

[33] Wang L, Amphlett G, Lambert JM, Blattler W, Zhang W. Structural characterization of a recombinant monoclonal antibody by electrospray time-of-flight mass spectrometry. *Pharm Res* 2005;22:1338–1349.

[34] Lyubarskaya Y, Houde D, Woodard J, Murphy D, Mhatre R. Analysis of recombinant monoclonal antibody isoforms by electrospray ionization mass spectrometry as a strategy for streamlining characterization of recombinant monoclonal antibody charge heterogeneity. *Anal Biochem* 2006;348:24–39.

5

CASE STUDY ON DEFINITION OF PROCESS DESIGN SPACE FOR A MICROBIAL FERMENTATION STEP

Pim van Hoek, Jean Harms, Xiangyang Wang, and Anurag S. Rathore

5.1 INTRODUCTION

The concept of "design space" has received a lot of attention in the biotech community [1]. It is defined in the "ICH Q8 Pharmaceutical Development" guideline as "The multidimensional combination and interaction of input variables (for example, material attributes) and process parameters that have been demonstrated to provide assurance of quality." The guideline describes best practices for pharmaceutical product development and emphasizes building quality into products by design, that is, designing processes with a low level of risk for failure to achieve the desired product quality attributes. This scientific approach to meet specific objectives, also known as the "Quality by Design" concept, encourages demonstration of a higher degree of understanding of the manufacturing process. This results in robust processes with better controls and less variability and greater confidence of assurance of quality, which translates into increased flexibilities in manufacturing and regulatory oversight [2]. The latter is elucidated in the following excerpt from the guideline: "Working within the design space is not considered

Quality by Design for Biopharmaceuticals, Edited by A. S. Rathore and R. Mhatre
Copyright © 2009 John Wiley & Sons, Inc.

as a change. Movement out of the design space is considered to be a change and would normally initiate a regulatory postapproval change process. Design space is proposed by the applicant and is subject to regulatory assessment and approval." The primary benefit of an approved design space is regulatory flexibility, most notably the potential to make process improvements within the design space with reduced regulatory oversight [3, 4].

Process characterization studies are performed primarily at laboratory scale with the purpose of defining the design space. Well-designed process characterization studies can serve as a foundation for a successful process validation, regulatory filing/approval, and subsequent manufacturing support over the life cycle of the product [4, 5]. In the last decade, these studies have been widely used in the biotech industry as a necessary precursor for setting performance acceptance criteria for subsequent successful process validation at manufacturing scale. This is demonstrated by the recent publications on related topics. The stepwise approach to characterization of biotech processes consists of risk assessment, small-scale model (SSM) qualification, process characterization studies, and setting process validation acceptance criteria based on process characterization and historical process data [4, 5]. Failure modes and effects analysis (FMEA) has been proposed as a tool that provides a rational approach to evaluating a process and generating a ranked order of parameters requiring process characterization [6].

Several publications have addressed the topics and provides guidelines for performing small-scale modeling and qualification [4, 5]. Shukla et al. demonstrated the utility of a design of experiments (DOE) approach to performing process characterization of a metal-affinity chromatographic purification process for an Fc fusion protein [8]. A fractional factorial study was performed to examine key operating parameters and their interactions. The results of the DOE were subsequently used to design the worst case studies to examine the robustness of the process. More recently, publications have addressed the design and approach toward process characterization of cell culture [9] and ion exchange unit operations [10]. Kaltenbrunner presented an alternative screening method for identifying potential key operating parameters for an ion exchange chromatography process. The advantages of this method based on chromatographic theory over the commonly used fractional factorial screening method were discussed. Mollerup et al. [11] have recently published thermodynamic modeling of chromatographic separation of proteins. Their approach aimed at supporting a more in-depth understanding of the design and development that is necessary in the Quality by Design paradigm. They were able to successfully simulate the chromatographic step and demonstrate the model's usefulness in identifying aberrations in step performance. Finally, the statistical analysis approaches for setting process validation acceptance criteria have been published recently [12]. Different statistical approaches including mean $\pm 3 \times$ standard deviations, tolerance interval analysis, prediction profiler, and Monte Carlo simulation were compared and benefits and disadvantages of these methods discussed.

This chapter presents a stepwise approach to establishing the design space for a biotech product in the form of a case study. It also discusses how the design space can reduce process validation requirements and enhance regulatory flexibility.

5.2 APPROACH TOWARD PROCESS CHARACTERIZATION

The objective of process characterization studies is to evaluate the robustness of the process and to define the "design space" within which the process can operate and still perform in an acceptable fashion with respect to product quality and process consistency [4, 5]. Data generated in these studies serves as a basis for the following:

1. Categorization of parameters into critical (those that impact product quality), key (those that impact process consistency), and non-key (those that are neither critical nor key).
2. Identification of acceptable ranges (ARs) that define the design space.
3. Establishing acceptance criteria for process validation.

Technical information from the characterization study has become a regulatory expectation in recent years as an important precursor a pre-requisite for manufacturing process validation as well as for the long-term manufacturing support [13]. Since performing a characterization study at the manufacturing scale is not practically feasible due to cost of operation and limited availability of facility and equipment, small scale models that represent the performance of manufacturing scale process are usually employed in process characterization studies at the laboratory scale. The overall procedure of process characterization is shown in Fig. 5.1.

The general process flow diagram for the case study is shown in Fig. 5.2. In brief, the case study involves a recombinant human therapeutic protein that is produced in *Escherichia coli* grown in a complex medium under the control of a temperature-sensitive promoter. This protein is expressed intracellularly as insoluble inclusion bodies in high cell density fed-batch fermentations. This protein is expressed as intracellular, insoluble form and this allows removal of several fermentation process contaminants by centrifugation and washing of the inclusion bodies suspension during the downstream operation. The manufacturing process begins with an inoculum scale-up from a shake flask to a seed fermenter, which is then used to seed the production fermenter. Three phases occur during the production fermentation. After inoculation of the production fermenter, the culture grows in batch mode to a target optical density. By then, glucose in the fermentation broth is nearly depleted, and the fed-batch phase begins with the addition of Feed-1 solution. During the fed-batch phase, the agitation speed is fixed. The Feed-1 solution is added at an exponential rate to provide sufficient nutrients for exponential cell growth until the target OD is reached. Once the cell density reaches this point, Feed-1 addition is discontinued, and the induction phase begins with changes in temperature to activate the temperature-sensitive promoter and induce protein expression. During the induction phase, the Feed-2 solution is added at a fixed rate to provide additional amino acids for product synthesis. After a fixed period, the induction phase ends. The fermentation broth is then chilled. Subsequently, microfiltration/diafiltration and disc-stack centrifugation steps are performed to generate cell paste intermediate (CPI) that is stored frozen until the start of the purification process up to the final formulated drug product. In the following sections, we use this case study to

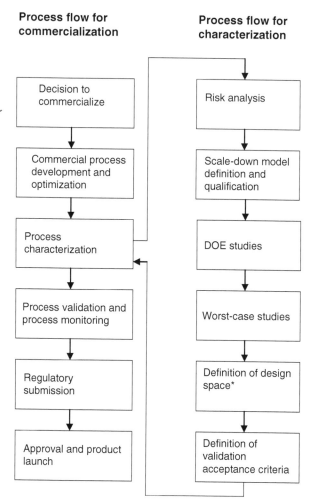

Figure 5.1. Overall process flow chart for commercialization of a biological product. The location of the process characterization subprocess is indicated. *After limits for critical product quality attributes are defined.

elucidate how the above-mentioned approach to process characterization can be followed for defining process design space.

5.3 RISK ANALYSIS

FMEA is a risk assessment tool widely used in the biotech industry for its underlying rigorous methodology for evaluating, prioritizing, and documenting potential failures of process and product performance [6]. In this case study, FMEA was used as a tool for prioritizing operational parameters for process characterization studies based on process knowledge. This approach involves assessment of operational parameters by

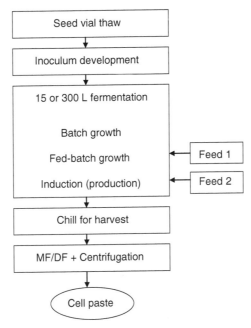

Figure 5.2. General process flow diagram for the case study.

an interdisciplinary team consisting of representatives from process/analytical development and manufacturing. The parameters are assessed with respect to the severity of operational parameter excursion (S), frequency of occurrence of the excursion (O), and ease/difficulty of detection of the excursion (D) [5]. Then, a risk priority number (RPN $= S \times O \times D$) score is calculated and used in the prioritization process.

In the case study shown in Fig. 5.3 and based on the review of the RPN scores, parameters exhibiting an RPN score >20 were considered a risk high enough to merit further characterization. These included the inoculum density for the seed fermenter step; acid/base control and dissolved oxygen (DO) for the growth phase of production fermenter; and optical density for induction and dissolved oxygen for the production phase of production fermenter. Some parameters with lower RPN scores, for example, pH and temperature during growth and induction phase, were included for characterization due to their high severity scores so as to generate process understanding for future process improvement efforts.

5.4 SMALL-SCALE MODEL DEVELOPMENT AND QUALIFICATION

A qualified small scale model is critical to process characterization, process-fit studies, manufacturing troubleshooting, and process improvement studies [4, 5, 7]. An acceptable small scale model needs to not only represent the large-scale performance at operating set point but also in the operating range and beyond (Fig. 5.4). This ensures that the results generated from the process characterization studies are applicable to the process at manufacturing scale. Figure 5.5 illustrates an approach toward small-scale

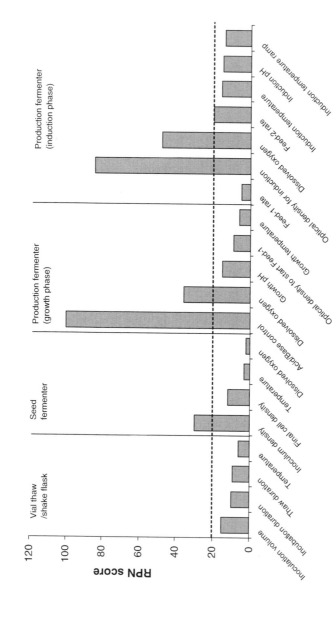

Figure 5.3. Pareto chart of RPN scores calculated based on a risk-assessment for operational parameters from different phases during the production process in the case study. *Note*: An RPN score >20, as indicated by the dotted line, was considered a risk high enough to merit further characterization.

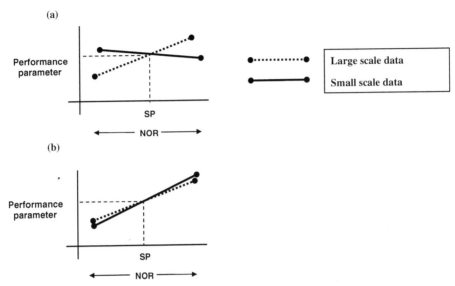

Figure 5.4. A successful small-scale model mimics not only the large-scale process performance at SP operating conditions but also that observed at the edges of (and preferably beyond). Normal Operating Range (NOR). A. Small-scale model mimics large-scale performance at center point but deviates if the process is run elsewhere in NOR. B. Small-scale model mimics large-scale performance not only at center point but also in the NOR.

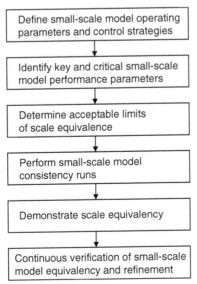

Figure 5.5. Small-scale model development and qualification activities.

TABLE 5.1. Operational Parameter Classification Based on Scale Dependency (Assuming Equivalent Bioreactor Design)

	Scale Factor
Scale-dependent parameters	
Batch volume (L)	100
Feed-1 rate (g/h)	100
Feed-2 rate (g/h)	100
Agitation (RPM)	As per P/V
Scale-independent parameters	
Seed density	1
Growth temperature (°C)	1
Induction temperature (°C)	1
pH set point	1
Induction duration (h)	1
Backpressure (psig)	1

modeling and qualification. The approach involves choosing a small-scale modeling strategy, setting predetermined qualification criteria, performing small-scale runs, and analyzing data to ascertain that qualification criteria were successfully met. Model verification and refinement should occur over the life cycle of the product to incorporate process changes made for process improvements and/or troubleshooting.

For the case study under consideration, the aspect ratios of vessel geometry with respect to the diameters of the tank and impeller, liquid height, baffle width and number, and sparger and impeller type were kept similar between scales. The operational parameters that were selected and prioritized from the FMEA were divided into scale-dependent and scale-independent operational parameters (Table 5.1). While the scale-independent parameters were kept identical between scales, scale-dependent parameters were scaled down linearly based on initial culture batch volume with the exception of the agitation speed for which scaling was based on maintaining a constant power per unit volume (P/V). Table 5.2 lists the various scaling strategies that can be used depending on the application under consideration [15]. The P/V strategy is aimed at providing constant k_La and thus similar gas transfer characteristics for controlling oxygen supply when similar equipment geometry and sparger design are assumed. Since the characteristic mixing times for a typical bench scale fermenter system are much shorter (measured in seconds) than that for a manufacturing scale system (typically on the scale of minutes) at similar P/V values, scaling down based on mixing time would require a significant reduction in power per volume as described in Table 5.2, resulting in poor gas transfer characteristics. The microbial cells in the case study are not sensitive to "shear" stress. According to the literature, high local energy dissipation rates rather than tip speed are a more reliable indicator of "shear" stress to cells [15]. Therefore, scaling agitation based on tip speed relating to managing shear stress is not relevant in our case. Finally, scaling down based on a constant Reynolds (Re) number does not apply when operating in the turbulent flow regime in baffled fermenters. In this flow regime, the power number of the impeller, which affects the delivered power per volume, becomes independent of Re number [16].

TABLE 5.2. Scale-Down Implications of Different Scaling Criteria Based on Agitation Speed at Constant P/V

Basis for Scaling Criteria	Consideration	Equation	Scaling Relationship	Scale-Down Relationship Assuming Equal P/V and Aspect Ratios ($N \propto D^{-2/3}$)
Power/volume (P/V) at constant V_s	Constant $k_L a$	$Po_g \cdot N^3 \cdot D_i^5 / V$	$\propto N^3 \cdot D_i^2$	Constant
Tip speed	Similar "shear stress"	$\pi \cdot N \cdot D_i$	$\propto N \cdot D_i$	$\propto D_i^{1/3}$ (lower)
Mixing time (θ_m)	Similar characteristic mixing time	$5.9 T^{-2/3} \cdot (P/V)^{-1/3}$ $\cdot (D_i/T)^{-1/3}$ for $H = T$	$\propto N^3 \cdot D_i^2, T^{2/3},$ $(D_i/T)^{-1/3}$ for $H = T \propto (H/D_i)^{2.5}$ for $H > T$	$\propto T^{2/3}$ (lower)
Reynolds number (Re)	Similar turbulent flow patterns	$N \cdot D_i^2 \cdot \rho / \mu$	$\propto N \cdot D_i^2$	$\propto D_i^{4/3}$ (lower) not relevant when remaining in turbulent flow regime

Source: Based on [16].

V_s, superficial gas velocity; V, volume; Po_g, impeller power number under sparging conditions; N, agitation rate (rev s^{-1}); D_i, impeller diameter; T, tank diameter; H, liquid height; ρ, liquid density; η, viscosity (Pa·s); P, power (watts).

TABLE 5.3. Fermentation SSM Qualification Against Manufacturing Performance Criteria

	Product Yield (g)	Total Elapsed Time to Induction (h)	Biomass (OD)
SSM Mean ± SD	100 ± 3	26.3 ± 0.3	100 ± 2
SSM interval (corrected for sample frequency)	94.4–106 (94.4–98.0)	24.8–27.0 (27.76–28.24)	89.1–109 (90.2–97.6)
Manufacturing data tolerance interval ($\alpha = 0.95$, $p = 0.95$)	86.5–135	26.6–29.9	81.3–113.8

Product yield and biomass data have been normalized against the SSM mean performance. SD, standard deviation.

A small-scale model was developed and qualified on the basis of manufacturing-scale data for this case study. The small-scale model is deemed qualified if the performance parameters fall within the tolerance interval bounds determined from manufacturing data. Table 5.1 shows the key operational parameters of the small-scale model. The growth performance in the small-scale model, expressed here as the total elapsed time to induction, initially did not meet the acceptance criteria based on manufacturing scale performance although all other qualification criteria were achieved (Table 5.3). As seen in Fig. 5.6, biomass growth and the corresponding dissolved oxygen profile at the manufacturing scale lag behind that in the small-scale model from the early stages of the fed-batch phase onward. The root cause of this discrepancy was the relative high sampling frequency at bench scale. When the sample frequency was accounted for the small-scale model met all the acceptance criteria of the manufacturing scale production (Table 5.3). Therefore, the small-scale model was deemed as qualified.

5.5 DESIGN OF EXPERIMENT STUDIES

These studies are performed on the basis of OPs identified and prioritized by FMEA and using qualified small scale models. The characterization ranges (CRs) were set at three times the normal operating range (NOR). A variety of DOE studies can be planned depending on the number of parameters that need to be examined and the resolution required for the study [4, 8, 10]. The approaches vary in their resolution and the number of experiments required for a given number of parameters to be studied. For our case, fractional factorial and full factorial studies conducted were deemed appropriate. The fractional factorial DOE study was designed such that the main effects could be distinguished from all two-factor interactions for those parameters expected to have the most effect on process performance. Each experiment was performed at least in duplicate, and the DOE designs contained multiple center points used for independent estimation of pure error as well as identification of potential quadratic curvature in the data. All experiments were carefully monitored to ensure control of the operation parameter(s) within the CR.

Data from fractional and full factorial DOE design studies were analyzed by analysis of variance (ANOVA) to understand main effects of operational parameters on process

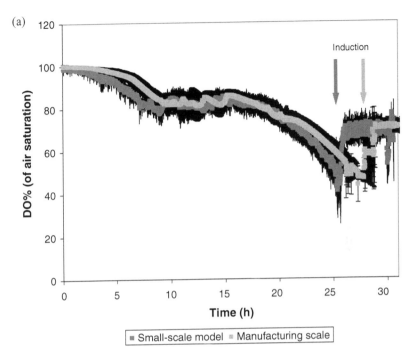

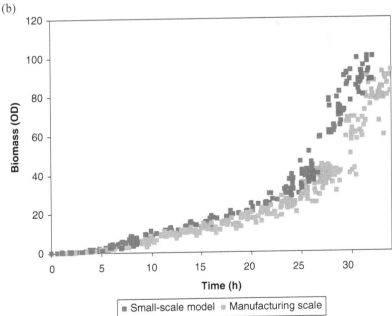

Figure 5.6. Comparison of DO% and growth profiles as a function time between the small-scale model and the manufacturing scale process. *Note*: Biomass data have been normalized against the mean final biomass density in the small-scale model under set point operating conditions.

performance. A two-sided confidence level (α) was set at 0.05 for this analysis. This means that when the p value is <0.05, the effect of an operational parameter on the related performance parameter is recognized as statistically significant. However, when p is >0.05, the impact of operational parameter on the performance parameter is not statistically significant. For the case study, analysis of product yield, product quality 1, and product quality 2 data was performed. Prediction profiles based on regression analysis for each operational parameter were generated and presented in Fig. 5.7. They illustrate how each operational parameter affects each performance parameter within the characterization range. For example, an increase in the temperature during growth-phase results in a decrease in product yield and an increase in both product quality attributes (Fig. 5.7a). Figure 5.7b and c illustrates similar trends for the induction phase of the production fermentation step.

While the prediction profiles are useful in illustrating the trends, review of the scaled estimates provides a more quantitative comparison. Figures 5.8–5.10 show the scaled estimates for the three data sets presented in Fig. 5.7. Figure 5.11 further illustrates the interaction effects that are observed in the process characterization studies. As mentioned earlier, the p-value denotes the statistical significance of the effect while the scaled estimate represents the magnitude of the effect. For process robustness, effects were examined that are both significant in magnitude and statistically significant. By defining >10% change from the mean for the scaled effect estimate to be considered significant in magnitude, the following conclusions can be made upon review of the data in Figs. 5.8–5.10. For the growth phase of the production fermentation (Fig. 5.8), effect of temperature on product yield; effect of temperature, feed-1, and pH on product quality 1; and effect of temperature and pH on product quality 2 are both statistically significant and of significant magnitude. Further, first-order interactions between feed-1 and pH show a statistically significant effect on both product yield and product quality 1 (Fig. 5.11a and b), but only for product quality 1 is this interaction effect also significant in magnitude. For the induction phase of the production fermentation (Figs. 5.9 and 5.10), effect of dissolved oxygen on product quality 1 and pH on product quality 2 are both statistically significant and of significant magnitude. Further, first-order interactions between feed-2 and pH have a significant effect on product quality 2 (Fig. 5.11c), in contrast to the interaction effect between Feed-2 and dissolved oxygen that only is statistically significant.

5.6 WORST CASE STUDIES

A worst case study was conducted to demonstrate process robustness and to determine whether process consistency and product quality performance would still be acceptable when operating each unit at the edge of the normal operating range. On the basis of the identification of operating parameters affecting product quality and process consistency (including product yield), a set of best case and worst case cell paste product intermediate was produced that exhibited either a high product titer and high purity or a low product titer and low purity (Table 5.4). The worst case product intermediate was further purified over an anion exchange (AEX) column. It was demonstrated that even under worst case operating conditions the acceptance criteria for product quality 2 impurity levels after the AEX purification step was never exceeded, assuring acceptable product quality of the

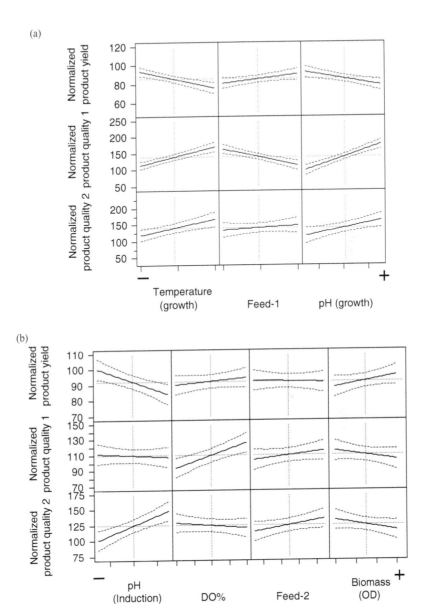

Figure 5.7. Prediction profiler showing main effects of growth (a) and induction phase (b, c) operational parameters on product yield and product quality performance. The solid lines represent the 95% confidence interval for the true mean. *Note*: Data has been normalized against the average small-scale model performance at set point operating conditions. The symbols − and + indicate the low and high end of the characterization range, respectively.

(c)

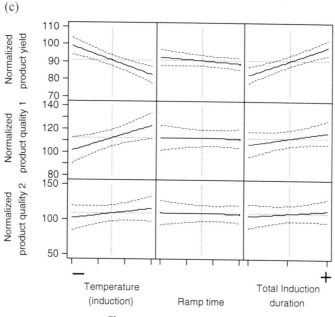

Figure 5.7. (Continued)

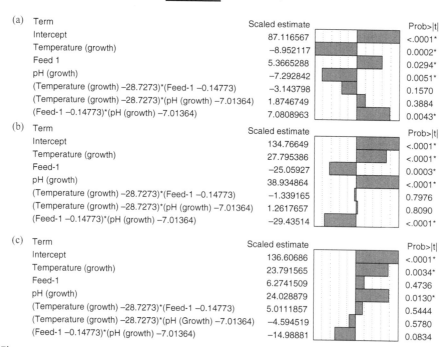

(a)

Term	Scaled estimate		Prob>|t|
Intercept	87.116567		<.0001*
Temperature (growth)	−8.952117		0.0002*
Feed 1	5.3665288		0.0294*
pH (growth)	−7.292842		0.0051*
(Temperature (growth) −28.7273)*(Feed-1 −0.14773)	−3.143798		0.1570
(Temperature (growth) −28.7273)*(pH (growth) −7.01364)	1.8746749		0.3884
(Feed-1 −0.14773)*(pH (growth) −7.01364)	7.0808963		0.0043*

(b)

Term	Scaled estimate		Prob>|t|
Intercept	134.76649		<.0001*
Temperature (growth)	27.795386		<.0001*
Feed-1	−25.05927		0.0003*
pH (growth)	38.934864		<.0001*
(Temperature (growth) −28.7273)*(Feed-1 −0.14773)	−1.339165		0.7976
(Temperature (growth) −28.7273)*(pH (growth) −7.01364)	1.2617657		0.8090
(Feed-1 −0.14773)*(pH (growth) −7.01364)	−29.43514		<.0001*

(c)

Term	Scaled estimate		Prob>|t|
Intercept	136.60686		<.0001*
Temperature (growth)	23.791565		0.0034*
Feed-1	6.2741509		0.4736
pH (growth)	24.028879		0.0130*
(Temperature (growth) −28.7273)*(Feed-1 −0.14773)	5.0111857		0.5444
(Temperature (growth) −28.7273)*(pH (Growth) −7.01364)	−4.594519		0.5780
(Feed-1 −0.14773)*(pH (growth) −7.01364)	−14.98881		0.0834

Figure 5.8. Scaled estimates of main effects of growth phase operational parameters on product yield (a), product quality 1 (b), and product quality 2 (c) performance from a full factorial DOE study. *Note*: Data have been normalized against the average small-scale model performance at set point operating conditions.

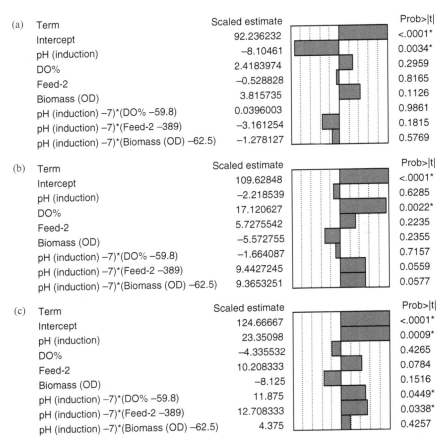

Figure 5.9. Scaled estimates of main effects of induction-phase operational parameters on product yield (a), product quality 1 (b), and product quality 2 (c) performance from a fractional factorial DOE study. *Note:* Data have been normalized against the average small-scale model performance at set point operating conditions.

final purified bulk product. This supports the demonstration of process robustness even under worst case operating conditions.

5.7 DEFINITION OF DESIGN SPACE

Figure 5.12 illustrates the concept of design space. The characterized space is examined in the process characterization studies, and the outcome of the studies defines the acceptable range for the operational parameters. For robust processes, it is expected that the normal operating ranges (NORs) for these parameters will be well nested within the respective acceptable range (AR). The combination of the ARs defines the design space for a process. It means that variability in the operational parameters within the AR would still lead to process performance meeting expectations. The expectations are outlined by the process validation acceptance criteria (PVAC) for the critical quality attributes of the product. As mentioned

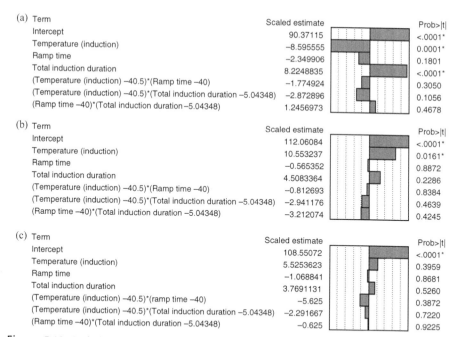

Figure 5.10. Scaled estimates of main effects of induction-phase operational parameters on product yield (a), product quality 1 (b), and product quality 2 (c) performance from a full factorial DOE study. *Note:* Data have been normalized against the average small-scale model performance at set point operating conditions.

earlier and illustrated in Fig. 5.12, implementation of the design space concept should allow the manufacturers to make changes in operating conditions without getting preapproval from the regulatory bodies as far as the changes are still within the approved design space.

Classification of performance parameters is primarily based on process development studies from which it is determined whether a performance parameter is a useful measure of process consistency (key) and/or product quality (critical) for a particular unit operation. Of all the variability in operational parameters that was examined, none resulted in unacceptable product quality as ascertained by the results from the worst case study shown in Table 5.4. Hence, for the fermentation step in the case study it was determined that there were no critical performance parameters. Step yield, product quality 1, and product quality 2 were identified as key performance parameters (KPPs).

Process characterization data are also used to classify operational parameters. This is done based on the effect of an operational parameter on the key and critical performance parameters of the process. As illustrated in Fig. 5.13, if variability of an operational parameter will result in a critical performance parameter (CPP) to deviate outside the respective PVAC, then it would be identified as a critical operational parameter. If the variability in the operational parameter would not result in a CPP to fail its PVAC but does have a significant impact on the CPP or KPP, the parameter is identified as a key operational parameter. Parameters that do not significantly impact CPP or KPP are identified as nonkey operational parameters. For the case study, and as

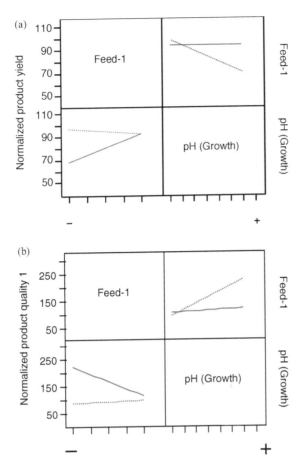

Figure 5.11. Statistical significant interaction effects of operational parameters on product yield or product quality performance during the growth (a, b) and induction phase (c). *Note:* The dotted and solid lines represent the low and high value, respectively, of the operational parameter shown on the right axis. Performance parameter values are shown on the left axis as a function of the operational parameter shown above or below the individual graphs. Data have been normalized against the average small-scale model performance at set point operating conditions. The symbols − and + indicate the low and high end of the characterization range, respectively.

mentioned earlier, the worst case combination of operational parameters resulted in acceptable product quality when processed through the process. Thus, none of the operational parameters was identified as critical. Table 5.5 presents a list of the key operational parameters that were identified. Temperature, Feed-1, and pH during growth phase and dissolved oxygen, Feed-2, and pH during induction phase were identified as key operational parameters. Table 5.5 also shows the acceptable ranges for each of these parameters. Together these acceptable ranges define the process design space.

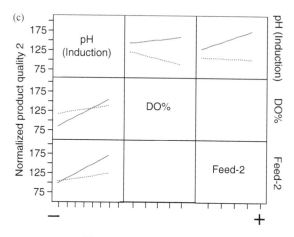

Figure 5.11. (*continued*).

TABLE 5.4. Worst and Best Case Operation Conditions and Performance Characteristics During the Production Stage in the Case Study

Operational Parameter	Worst Case		Set Point		Best Case	
	Growth Phase	Production Phase	Growth Phase	Production Phase	Growth Phase	Production Phase
DO%	+	+	0	0	−	−
Temperature	+	+	0	0	−	−
Feed-1	−	N/A	0	N/A	−	N/A
pH	+	+	0	0	−	−
Induction duration	N/A	−	N/A	0	N/A	+
Performance Parameter	Worst Case		Set Point		Best Case	
Product yield	57 ± 5		100 ± 4		114 ± 4	
Product quality 1	198 ± 9		100 ± 18		76 ± 2	
Product quality 1 (post-AEX)	226 ± 9		110 ± 14		n.d.	
Product quality 2	93 ± 13		100 ± 2		87 ± 7	
Product quality 2 (post-AEX)	550 ± 17		417 ± 93		n.d.	

Data have been normalized against the average bench scale performance at set point operating conditions at the end of the production phase. N/A, not applicable; n.d., not determined; AEX, anion exchange chromatography. The symbols −, 0, and + indicate the low end, the set point, and the high end of the normal operating range, respectively.

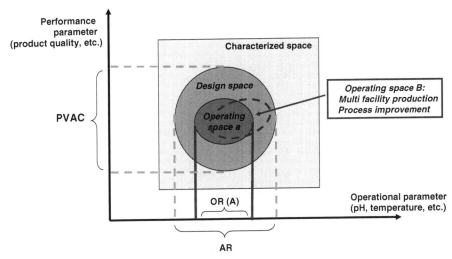

Figure 5.12. Design space philosophy enables flexibilities in manufacturing and regulatory filing. AR, acceptable range; PVAC, process validation acceptance criteria; OR, operating range.

5.8 DEFINITION OF VALIDATION ACCEPTANCE LIMITS

Process validation provides documented evidence with a high degree of assurance that a specific process at manufacturing scale will consistently produce a product meeting its

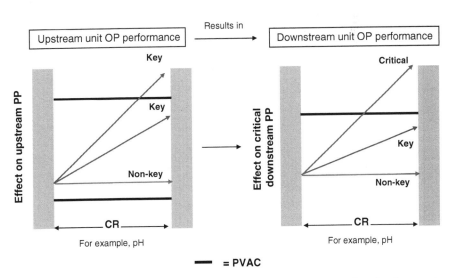

Figure 5.13. Classification of OPs based on performance data derived within the characterization range with respect to key and critical performance parameters. *Note:* The gray areas represent the noncharacterized design space for the operational parameter under investigation. PVAC, process validation acceptance criteria.

TABLE 5.5. PC-Based Classification of Operational Parameters and Setting of Acceptable Operating Ranges

Operation Parameter	NOR	AR	CR	Parameter Class
Growth-phase temperature (°C)	±2	±2	±2	Key
Feed-1 (h^{-1})	N/A	0.10 – 0.15	0.10 – 0.20	Key
pH (growth)	±0.1	±0.1	±0.3	Key
pH (induction)	±0.1	±0.1	±0.3	Key
Feed-2 (g/h)	±10	± 40	±40	Key
DO (% of air saturation)	>50%	>25%	25–100	Key

NOR, normal operating range; AR, acceptable range; CR, characterization range.

predetermined specifications and quality attributes [17]. It is a regulatory expectation that valid in-process acceptance criteria for characteristics related to process consistency and product quality are defined that are consistent with the final specifications of the drug product [18]. Acceptable variability in product quality and process performance attributes is established based on knowledge from clinical exposure of the product, or from other similar products, and general scientific understanding about the molecule. Process validation acceptance criteria should be derived from appropriate statistical procedures based on data from extensive process characterization studies and historical process data that demonstrated acceptable process performance and product quality. Prior to initiation of process validation, these numerical limits should be set on the basis of the process performance observed within the wider acceptable range (in contrast to the normal operating range) of operational parameters. When these PVAC do get exceeded, this signals a significant departure from acceptable operating conditions that could result in a failed validation lot. Setting PVAC inappropriately results in either inadequate measure of process performance (when too wide) or invalidating acceptable manufacturing runs (when too narrow).

When extensive historical and process characterization data are available, as in this case study, and can be modeled with sufficient accuracy, Monte Carlo simulation is the appropriate statistical method for setting PVAC [12, 19]. This case study illustrates the approach for calculating PVAC for the product quality 1 performance parameter as an example. First, the entire data set containing data from bench scale and pilot scale experiments was used to fit a regression model using statistical software. For fermentations not run at SP (i.e., at the edges of the characterization or normal operating ranges), the operating parameters including the operating scale were coded to the operating range for all the operating parameters evaluated in the characterization study. Initially, all operating parameters were considered in the model. Second, through step-wise backward regression model building a final model was constructed that contained the highest R^2 with four statistically significant operational parameters at probabilities of $p \leq 0.05$ (Figs. 5.14a and b). The residual distribution was evaluated for normality in order to establish whether the regression model was appropriate for further analysis.

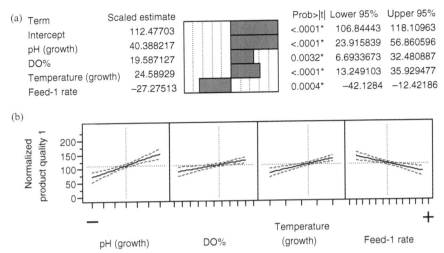

(a)

Term	Scaled estimate		Prob>|t|	Lower 95%	Upper 95%
Intercept	112.47703		<.0001*	106.84443	118.10963
pH (growth)	40.388217		<.0001*	23.915839	56.860596
DO%	19.587127		0.0032*	6.6933673	32.480887
Temperature (growth)	24.58929		<.0001*	13.249103	35.929477
Feed-1 rate	−27.27513		0.0004*	−42.1284	−12.42186

(b)

Figure 5.14. Building a regression model for product quality 1 based on the acceptable operating ranges. *Note:* Data have been normalized against the average small-scale model performance at set point operating conditions. The symbols − and + indicate the low and high end of the operating range, respectively.

Next, a Monte Carlo simulation was performed in which the ranges of statistically significant operating parameters were varied either triangular or uniformly within the acceptable operating ranges (Fig. 5.15a). A triangular distribution for the Feed-1 parameter was selected to better represent the expected operational variation and control tolerance around the target SP. Random noise was applied during the simulation that is derived from the root mean square error associated with the model prediction. Fourth, two-sided tolerance intervals were calculated based on the simulated performance data ($n = 10,000$) calculated to contain 99% of the population with 95% confidence (Fig. 5.15b). The calculated performance output adhered to a normal distribution. For generation of the initial model, all possible conditions (including interaction of multiple operational parameters within the acceptable operating range) were considered with equal weight or chance of occurrence. The Monte Carlo simulation appropriately assigns a higher probability to runs performed at SP of the operating ranges and lower probability to runs performed at the edges of the ranges.

The PVAC calculated by Monte Carlo simulation are presented in Fig. 5.16. They represent the tolerance interval describing 99% of the future data with 95% confidence. As shown, the process performance at SP operating conditions falls well within the PVAC. In addition, the observation that the worst and best case experiments within normal operating conditions fall also within the PVAC for product quality 1 further demonstrates the robustness of the process. It should be pointed out that data from the growth PC study that exceed PVAC were from experiments performed under extreme conditions. These resulted in unacceptably high product quality 1 levels and thus, they are expected to fall outside the proposed process validation acceptance criteria.

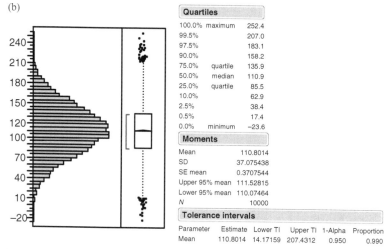

(a) Factors	Distribution	Type
Scale	Fixed	
pH (growth)	Random	Uniform
DO%	Random	Uniform
Feed-1 rate	Random	Triangular
Temperature (growth)	Random	Uniform
N runs	10000	
Random noise	S D: 31.5	

(b)

Quartiles

100.0%	maximum	252.4
99.5%		207.0
97.5%		183.1
90.0%		158.2
75.0%	quartile	135.9
50.0%	median	110.9
25.0%	quartile	85.5
10.0%		62.9
2.5%		38.4
0.5%		17.4
0.0%	minimum	−23.6

Moments

Mean	110.8014
SD	37.075438
SE mean	0.3707544
Upper 95% mean	111.52815
Lower 95% mean	110.07464
N	10000

Tolerance intervals

Parameter	Estimate	Lower TI	Upper TI	1-Alpha	Proportion
Mean	110.8014	14.17159	207.4312	0.950	0.990

<u>Figure 5.15.</u> Monte Carlo simulation parameters (a) and statistics (b) for the calculation of PVAC for product quality 1. *Note:* Data have been normalized against the average small-scale model performance at set point operating conditions.

5.9 REGULATORY FILING, PROCESS MONITORING, AND POSTAPPROVAL CHANGES

The design space in the form of the acceptable ranges for the key and critical operational parameters along with the process validation acceptance criteria should be included in the filing. After the product is approved, process monitoring should be performed to demonstrate that product quality and process performance attributes are within the filed design space. As mentioned earlier, the primary benefit of the design space concept is the regulatory flexibility. Process changes within the design space should require less regulatory review or approval. Therefore, process improvements during the product life cycle with regard to process consistency and throughput can be made with reduced postapproval submissions. However, as stated in ICH Q8, "The degree of regulatory

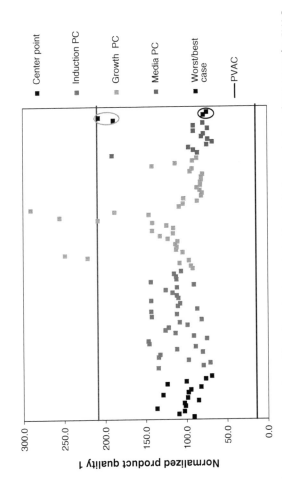

Figure 5.16. Projection of bench scale characterization data for product quality 1 onto the PVAC derived from Monte Carlo simulation. *Note:* The performance data of individual characterization studies (performed outside the normal operating range) is color coded. Data have been normalized against the average small-scale model performance at set point operating conditions. The best case performance within the normal operating range is represented by a black circle, while that for worst case performance is represented by a grey circle. Both circles contain results from two independent duplicate experiments. (See the insert for color representation of this figure.)

Legend:
- Center point
- Induction PC
- Growth PC
- Media PC
- Worst/best case
- PVAC

flexibility is predicated on the level of relevant scientific knowledge provided [1]." Design space and regulatory flexibility should not be confused with relaxed process control or increased process variability. Regardless of how wide the design space is, process consistency is still the goal for commercial manufacturing. Therefore, the manufacturing process would be performed in a reasonably narrow operating space based on equipment capability and for maintaining process consistency. Excursions outside the operating space would indicate unexpected process drift and initiate appropriate investigations into the cause of the deviation and subsequent corrective action. As long as operating parameters remain within the design space, however, product release should not be in jeopardy. As manufacturing experience grows and opportunities for process improvement are identified, the operating space could be revised within the design space without the need for postapproval submission. In addition, process understanding gained from process monitoring can also be used for future changes to the design space. Such changes should be evaluated against the need for further characterization and/or revalidation and subsequent filing.

ACKNOWLEDGMENT

The authors would like to thank Duncan Low (Amgen), Yuan Xu (Novartis), Tony Mire Sluis (Amgen), for their critical reading of the manuscript and helpful suggestions.

REFERENCES

[1] ICH Guidance for Industry: Q8 Pharmaceutical Development. Rockville, MD: U.S. Food and Drug Administration. 2006. www.fda.gov/cber/gdlns/ichq8pharm.htm.

[2] Rathore AS, Winkle H. Quality by Design for Pharmaceuticals: Regulatory Perspective and Approach. *Nature Biotechnology* 2009;27:26–34.

[3] Rathore AS, Branning R, Cecchini D. Design space for biotech products. *BioPharm Int* 2007;36–40.

[4] Harms J, Wang X, Kim T, Yang J, Rathore AS. Defining Design Space for Biotech Products: Case Study of Pichia Pastoris Fermentation. *Biotechnol Prog.* 2008; 24(3):655–662.

[5] Seely J. Process characterization. In: Rathore AS, Sofer G, editors. *Process Validation in Manufacturing of Biopharmaceuticals*. Boca Raton: Taylor & Francis; 2005. p 31–68.

[6] Seely RJ, Haury J. Applications of failure modes and effects analysis to biotechnology manufacturing processes. In: Rathore AS, Sofer G, editors. *Process Validation in Manufacturing of Biopharmaceuticals*. Boca Raton: Taylor & Francis; 2005. p 13–30.

[7] Godavarti R, Petrone J, Robinson J, Wright R, Kelley BD. Scaled-down models for purification processes: approaches and applications. In: Rathore AS, Sofer G, editors. *Process Validation in Manufacturing of Biopharmaceuticals*. Boca Raton: Taylor & Francis; 2005. p 69–142.

[8] Shukla AA, Sorge L, Boldman J, Waugh S. Process characterization for metal affinity chromatography of an Fc fusion protein: a design-of-experiments approach. *Biotechnol Appl Biochem* 2001;34:71–80.

[9] Li F, Hashimura Y, Pendleton R, Harms J, Collins E, Lee B. A systematic approach for scaled-down model development and characterization of commercial cell culture processes. *Biotechnol Prog* 2006;22:696–703.

[10] Kaltenbrunner O, Giaverini O, Woehle D, Asenjo JA. Application of chromatographic theory for process characterization towards validation of an ion-exchange operation. *Biotechnol Bioeng* 2007;98:201–210.

[11] Mollerup JM, Hansen TB, Kidal S, Staby A. Quality by design: thermodynamic modeling of chromatographic separation of proteins. *J Chromatogr A* 2008;1177(2):200–206.

[12] Wang X, Germansderfer A, Harms J, Rathore AS. Using statistical analysis for setting process validation acceptance criteria for biotech products. *Biotech Prog* 2007;23:55–60.

[13] ICH Guidance for Industry: Q5E Comparability of Biotechnological/Biological Products Subject to Changes in Their Manufacturing Process. 2005. www.fda.gov/cber/gdlns/ich-compbio.htm.

[14] Schmidt FR. Optimization and scale up of industrial fermentation processes. *App Microbiol Biotechnol* 2005;65:425–435.

[15] Nienow AW. Reactor engineering in large scale animal cell culture. *Cytotechnology* 2006; 50:9–33.

[16] Hewitt CJ, Nienow AW. The scale-up of microbial batch and fed-batch fermentation processes. *Adv Appl Microbiol* 2007;62:105–135.

[17] U.S. Food and Drug Administration. Guideline on General Principles of Process Validation. Rockville, MD: U.S. Food and Drug Administration; 1987. Draft Guidance, November 2008. http://www.fda.gov/CDER/GUIDANCE/8019dft.pdf.

[18] U.S. Food and Drug Administration. Current Good Manufacturing Practice for Finished Pharmaceuticals. Code of Federal Regulations. Title 21, Vol. 4, Part 211. Washington, DC: U.S. Government Printing Office; 2005. www.fda.gov/cder/dmpq/cgmpregs.htm.

[19] Hammersley JM, Handscomb DC. *Monte Carlo Methods*. New York: John Wiley & Sons; 1964.

6

APPLICATION OF QbD PRINCIPLES TO TANGENTIAL FLOW FILTRATION OPERATIONS

Peter K. Watler and John Rozembersky

6.1 INTRODUCTION

The concepts of Quality by Design (QbD) and design space have been stressed in recent U.S. and international regulatory publications including "Pharmaceutical cGMPs for the twenty-first century" [1, 2], "Guidance for Industry—PAT" [3], "Quality Risk Management" [4], "Pharmaceutical Development" [5], "Annex to Pharmaceutical Development" [6], and "Standard Guide for Specification, Design, and Verification of Pharmaceutical and Biopharmaceutical Manufacturing Systems and Equipment" [7]. The concept of QbD is quite straightforward and as explained by Moheb Nasr of the FDA, it simply involves "doing the right thing," taking "a good scientific approach," and "using good science" [8]. In essence, QbD involves employing the fundamentals of the scientific approach and using common sense. The main benefit of the knowledge gained from employing QbD is to enhance "the quality of pharmaceuticals" [8]. Furthermore, the greater process understanding and optimal operation of unit operations "will result in cost benefits for the industry." Finally, with a detailed understanding of the process, firms will see that "regulatory flexibility is an added benefit"; however, as Nasr notes, this is not the main driver for initiating QbD efforts.

Quality by Design for Biopharmaceuticals, Edited by A. S. Rathore and R. Mhatre
Copyright © 2009 John Wiley & Sons, Inc.

Engineering correlations and modeling equations form the basis for the design of many unit operations and equipment systems in the processing industries. The use of such an approach relies on having an in-depth understanding of the system in terms of its fluid dynamics, physics, and chemistry. Such an understanding affords a knowledge of the principles of a system's operation and how and why the operating parameters relate to system performance. The mechanistic modeling approach thus affords a complete understanding of the operation of systems and facilitates definition of the design space by embracing the principles of Quality by Design. In fact, the FDA's PAT Team and Manufacturing Science Working Group recently noted that a "mechanistic understanding, as opposed to data derived from one-factor-at-time type of experiment or simple correlative information, provides a higher level of knowledge and an ability to generalize within certain constraints." A well-planned and thoughtful mechanistic approach to design invokes the principles of the scientific method and relies further on empirical verification. As Hutchins [9] stated, "merely collecting facts will (not) solve a problem of any kind. The facts are indispensable; they are not sufficient. To solve a problem, it is necessary to think" [9].

The tangential flow filtration (TFF) unit operation is particularly well suited to a mechanistic Quality by Design approach since the mass transfer fundamentals are well known and proven engineering principles and design equations have been developed. This scientific approach provides a mechanistic understanding of the hydrodynamic principles governing this separation process. The detailed knowledge gained by employing mechanistic design principles identifies the optimal operating conditions and defines the design space, thus embracing the concepts of Quality by Design. Mechanistic TFF design equations have a solid basis from principles of boundary layer flow, Fickian diffusion, Darcy pore flow, film theory, eddy diffusivity model, and dimensional analysis using heat transfer analogies to describe turbulent and laminar flow based on Reynolds, Schmidt, and Sherwood numbers. As a result of the strong theoretical basis in fluid dynamics and mass transport, several mechanistic models have been developed to predict flux in TFF systems as a function of operating parameters such as pressure, viscosity, temperature, concentration, velocity, diffusivity, density, osmotic pressure, length, and diameter [10–12].

6.1.1 How to Use This Chapter

As described above, many design models have been proposed to address the various industrial applications of TFF. Many publications provide a detailed derivation of the theoretical principles and are sufficiently comprehensive to apply the fundamentals to food, oil, gas, and water treatment systems. Such detailed treatments can make it difficult for the user to extract the key equations and approaches needed to design a specific system. This chapter will focus only on design approaches suitable for biotechnology TFF applications and will provide a mechanistic design approach tailored to protein and cellular separations. The chapter describes a series of sequential steps required to select, optimize, and specify a TFF operation. The mechanistic basis for each step will be described along with the key design equations and approaches. Employing such proven mechanistic design principles will fulfill a "Quality by Design" approach that fully describes the TFF unit operation and provides a fundamental understanding of the operation unique to the membrane, feed stream, and fluidics system. Each step will conclude with a summary of the key "QbD Principles" required for the specification and optimization of the TFF unit operation.

6.2 APPLICATIONS OF TFF IN BIOTECHNOLOGY

The tangential flow filtration unit operation encompasses microfiltration, ultrafiltration, nanofiltration, and reverse osmosis processes. All are pressure driven membrane process and are differentiated by the membrane that determines which species permeate and which are retained [10]. In biotechnology, microfiltration (suspended particles 0.1–5 μm, e.g., cells) and ultrafiltration (macromolecules 0.001–0.002 μm, e.g., proteins) are the two most relevant applications of TFF. Concentration and diafiltration are the two TFF operating modes used in biopharmaceutical production. In upstream applications, TFF can be used to recover the target molecule and remove impurities from perfusion reactors. It may also be an effective means to wash and harvest host cells or to recover excreted products from the fermentation broth. In downstream applications, TFF is used to concentrate the target molecule in preparation for a subsequent unit operation, such as chromatography, or to adjust the product concentration to that required by the final formulation. Diafiltration is used to remove unwanted species such as impurities from the fermentation process (media components, host cell proteins, proteases) and buffer excipients. In this manner, TFF is frequently employed to condition a feed stream to the appropriate concentration, ionic strength, pH, and buffer components needed for subsequent unit operations. TFF is often the unit operation of choice for such applications because of its mechanical simplicity, robustness, and relatively low cost. However, an improperly designed and scaled TFF operation will be plagued with membrane fouling, low product recovery, long processing times, excessive buffer usage, frequent membrane replacement, and inconsistent operation. Such problems can be avoided by applying Quality by Design principles to understand and specify the TFF unit operation.

6.3 TANGENTIAL FLOW FILTRATION OPERATING PRINCIPLES

In tangential flow filtration (TFF), the primary driving force for permeation through the membrane is an external pressure gradient known as the transmembrane pressure (TMP) [13]. This pressure is much larger than the osmotic pressures across the membrane and reverses the flow of solvent and permeable solutes from the concentrated solution to the dilute solution. In microfiltration and ultrafiltration, the feed stream solute concentration is generally low and osmotic pressure is typically negligible. As a result, these processes are typically operated at relatively low external pressures in the range of 5–100 psi. Solute rejection also increases with the external pressure gradient (TMP), and thus TMP can also affect membrane selectivity.

As filtration proceeds, the retained molecules are unable to sufficiently diffuse away from the membrane surface. This is caused by the rapid transport of solvent through the

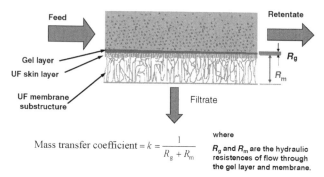

$$\text{Mass transfer coefficient} = k = \frac{1}{R_g + R_m}$$

where

R_g and R_m are the hydraulic resistances of flow through the gel layer and membrane.

Figure 6.1. Filtrate resistance to flow due to membrane and gel layer resistances.

membrane bringing nonpermeable solutes to the membrane surface where they accumulate. This phenomenon is known as concentration polarization, and it results in the formation of a thin solute layer at the membrane surface known as the "gel" layer [14]. The gel layer forms below the fluid boundary layer and creates an additional resistance to flow that together with the membrane resistance influences the mass transfer rate and the membrane selectivity (Fig. 6.1).

To counteract the adverse effect of concentration polarization, a high-velocity fluid stream is applied tangentially to the membrane surface. In some cases, screens are inserted in the membrane channels to promote turbulent flow, which reduces the fluid boundary layer compared to that of laminar flow. The tangential flow provides turbulence and shear that transport solutes from the membrane surface back to the bulk fluid stream. However, as the filtration proceeds and the retained solute concentration increases, the concentration polarization and the gel layer resistance increase. This results in an increase over time of the mass transfer coefficient, k, which lowers the filtrate flux rate (Fig. 6.2). This dynamic process means that TFF operations never operate in steady state, and as a result a precise understanding of the operating parameters

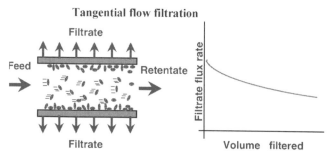

Figure 6.2. Tangential flow between membrane sheets transports solutes from membrane surface into bulk fluid stream. As solute concentration increase, gel layer resistance increases, reducing filtrate flux.

is especially critical to robust and consistent design. Hence, an understanding of both the membrane and the operating conditions are necessary in TFF to determine the solute separation. Fortunately, given the well-developed fundamentals of fluid flow and mass transfer, proven approaches have been developed and are summarized and presented here to provide a "Quality by Design" understanding of TFF system operation. These design approaches enable reliable specification of the transmembrane pressure driving force and the compensating tangential flow to optimally balance the complex and competing mass transport effects seen with TFF.

6.4 TFF DESIGN OBJECTIVES

In designing a TFF operation, the major design objectives in order of importance are as follows:

1. Product recovery
2. Membrane recovery
3. Filtrate flux rate

Designing for product recovery ensures cost-effective operation and involves evaluating factors such as membrane selectivity, membrane adsorption, and shear degradation. Membrane recovery enables cost-effective reuse of membranes, reduces change-out time, and aids in lot-to-lot consistency. Optimizing filtrate flux rate minimizes the membrane area and operating time. Since the operating parameters impact each of the above objectives, they must be simultaneously addressed during the design process. A clear set of QbD principles is thus critical to the optimal and robust specification of the TFF process.

6.5 MEMBRANE SELECTION

The first step to designing a TFF step is to select the membrane that will give the best separation performance for a given feed stream. Cellulose acetate, polyvinylidene difluoride (PVDF), polysulfone (PS), and polyether sulfone (PES) are the most common membrane types of membrane media employed in biotechnology applications [15]. Asymmetric membranes generally offer the best performance and consist of a thin skin ($\sim$0.1–0.5 µm) that serves as the solute barrier on top of a polymeric substrate layer ($\sim$100–300 µm) that supports the thin skin [16]. The thin skin of the membrane largely determines the rejection of solutes in the feed stream. The composite structure of asymmetric membranes provides both excellent retention and permeate flux. While the membrane itself is the dominant factor in determining retention of solutes, retention also depends on several operating factors. For example, protein transport across an ultrafiltration membrane has been modeled as the transport of a colloidal particle through a liquid-filled pore [17]. Since the size and shape of a protein is influenced by its pI (isoelectric point), solvent, pH, and ionic strength, these factors will affect the retention

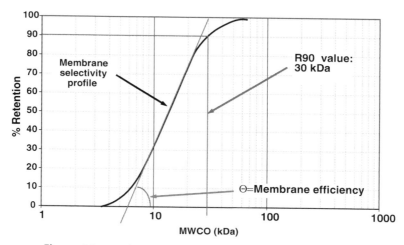

Figure 6.3. Membrane selectivity profile for determining MWCO.

of species. The membrane, concentration of the solute at the membrane surface, and to a lesser extent permeate flux and tangential flow velocity also influence the retention.

To aid in initial membrane selection, manufacturers provide molecular weight cutoff (MWCO) ratings for their membranes. The MWCO is often specified based on the R90 value that corresponds to a MW that is 90% retained by the membrane (Fig. 6.3). The MWCO is generally determined using dextran and is merely an approximation to actual solute retention.

As shown in Fig. 6.4, the shape of the membrane selectivity profile and MWCO rating varies significantly between membrane manufacturers. The more vertical the profile, the better the performance since more of the unwanted lower MW species

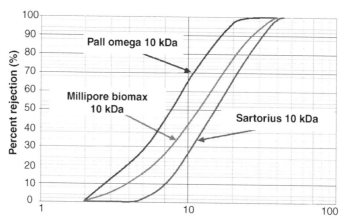

Figure 6.4. Membrane selectivity profiles for 10 kDa PES membrane from three manufacturers.

will be removed. For initial screening, the membrane should be selected with a MWCO value three to five times lower than the molecular weight of the species to be retained [18].

Species retention can be quantified by the rejection coefficient, R, [19]

$$R = 1 - \frac{C_p}{C_r}$$

where C_p = solute concentration in permeate and C_r = solute concentration in retentate.

The overall rejection coefficient for a given solute feed solution depends on (1) the membrane retention and adsorption properties and (2) the gel layer properties. Hence, the overall rejection coefficient is the combined rejection coefficient of membrane and gel layer. The gel layer is dependent on the solute feed properties (concentration, diffusivity coefficient, temperature), membrane–solute interactions and the operating conditions; transmembrane pressure and cross-flow velocity (turbulence and shear) across the membrane. Because the gel layer properties are dynamic and constantly changing, steady state of the gel layer formation is required to accurately determine the overall rejection coefficient. This is accomplished by keeping the solute feed concentration and system operating conditions constant. The rejection coefficient can be determined experimentally by operating the TFF system with the permeate open and recycling back to the retentate vessel. After the system comes to steady state, the concentration of the solute of interest is measured in the permeate and retentate streams. The rejection coefficient is then calculated using the above equation.

To minimize product loss to the permeate, especially during diafiltration, a membrane should provide a rejection coefficient $R > 0.99$. For example, at $R = 0.90$, 50% of the product will be lost to the permeate after just six diafiltration volumes. Conversely, to maximize the removal of unwanted low molecular weight species, the membrane should offer a low rejection coefficient of the contaminant, which will reduce the number of diafiltration volumes required to remove the contaminant.

In addition to product "leakage" through the membrane, product can also be lost due to nonspecific adsorption of the product to the membrane skin and substrate. Fortunately, advances in improving the hydrophilic nature of TFF membranes have generally removed membrane adsorption as a primary cause of product loss. Proteins can adsorb to membranes primarily through charge differences and hydrophobic interactions. This results in fouling of the membrane and has the effect of lower product recovery and degradation of filtrate flux and increases the cleaning difficulty membrane performance recovery [20]. Loss due to adsorption can be assessed during membrane selection. Recovery can be computed by performing a mass balance on the retentate and permeate.

6.5.1 QbD Principles for Membrane Selection

Maximize yield by selecting a membrane with high product retention and low adsorptive losses.

1. Measure the rejection coefficient, R, for the project and contaminants.
2. Calculate adsorptive losses from a mass balance on the product.

$$Y_{ads} = (C_{R0}V_{R0} - C_{Rf}V_{Rf} - C_{Pf}V_{Pf})/A$$

where Y_{ads} = adsorptive loss (mg/m^2); C_{R0}, C_{Rf}, C_{Pf} = initial and final retentate/permeate concentration; V_{R0}, V_{Rf}, V_{Pf} = initial and final retentate/permeate volume; A = membrane area.

3. Calculate the % product loss (or % contaminant removal) at a given diafiltration volume.

$$Y = 100 \times \left(1 - \left[e^{(V_p/V_0) \times (1-R)}\right]^{-1}\right)$$

where Y = % solute recovered in permeate, V_p = final permeate volume, and V_0 = initial feed volume.

6.6 TFF OPERATING PARAMETER DESIGN

To design quality into the TFF operation requires an understanding the dominant mass transfer principles governing permeate flux. There are four major operating parameters that can be controlled on a TFF system:

1. Transmembrane pressure
2. Cross-flow velocity that when normalized to the membrane area is termed the cross-flow flux (CFF)
3. Feed concentration
4. Temperature

Transmembrane pressure is the primary driving force for convective transport through the membrane pores and results in solvent flow from the feed to the permeate stream as shown in Fig. 6.5. The transmembrane pressure can be calculated from the following equation.

$$\text{TMP} = \frac{(P_{feed} - P_{permeate}) + (P_{retentate} - P_{permeate})}{2} = \frac{\text{TMP}_{feed} + \text{TMP}_{retentate}}{2}$$

The cross-flow flux is calculated as

$$\text{CFF} = \frac{Q_R}{A}$$

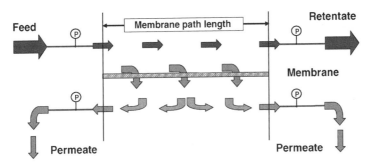

Figure 6.5. Schematic of hydraulic pressure and flow in tangential flow filtration.

where P_{feed} = retentate side inlet pressure, $P_{permeate}$ = permeate side pressure, $P_{retentate}$ = retentate side outlet pressure, Q_R = retentate flow rate, and A = membrane area.

Wheelwright [20] proposed a simplified mass transfer model based on the hydraulic resistances of the membrane and the gel layer. For a system where constant viscosity can be assumed, the governing equation for permeate flux becomes

$$J_{filtrate} = k \times TMP = \left[\frac{1}{R_g + R_m} \right] \times TMP$$

where k = mass transfer coefficient, TMP = transmembrane pressure; R_g = gel layer resistance, and R_m = membrane resistance.

As shown in Fig. 6.6 and described by Ladisch [21], there are three distinct regions where the transmembrane pressure and the cross-flow rate influence permeate flux to varying degrees. At lower transmembrane pressures, permeate flux is influenced by both membrane and gel layer resistances and is proportional to the pressure. In this region,

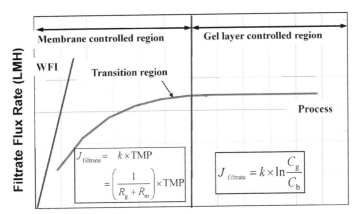

Figure 6.6. Effect of transmembrane pressure on permeate flux and the three TFF operating regions.

permeate flux is said to be membrane controlled and the flux is pressure dependent and is governed by the above equation.

At high transmembrane pressures, a significant gel layer of the rejected solutes accumulates on the membrane and restricts flow. At this point, mass transfer is no longer controlled by the operating parameters of cross-flow and transmembrane pressure. In this region, permeate flux is said to be gel layer controlled and the flux is independent of pressure. In this region, there is a balance between the rate at which solute is deposited at the membrane due to convection by the solvent flow through the membrane and diffusion of the solute from the membrane to the bulk liquid. Here, permeate flux is governed by the mass transfer coefficient and the relative solute concentration in the gel layer and bulk solution (feed concentration), as shown in the following equation for the film theory model that predicts that flux decreases exponentially with feed concentration

$$J_{\text{filtrate}} = k \times ln\frac{C_{G}}{C_{B}}$$

where C_{G} = gel layer concentration and C_{B} = bulk stream concentration.

The optimal operating point is in the transition region between the membrane controlled region and the gel layer controlled region. In this region, the beneficial effect of increasing transmembrane pressure on permeate flux becomes nonlinear, yet flux is still controlled by TMP. Here, both transmembrane pressure and cross-flow flux are important. Cross-flow provides turbulence and shear that improve the mass transfer of solutes from the membrane surface back to the bulk liquid. The agitation and mixing provided by the cross-flow turbulence effectively sweeps accumulated solute from the membrane and reduces the gel layer thickness by controlling concentration polarization. Increasing cross-flow initially has a linear effect on reducing the gel layer thickness and moves the gel layer controlled region to higher transmembrane pressure as shown in Fig. 6.7. In the transition region, cross-flow affects filtrate flux by extending the

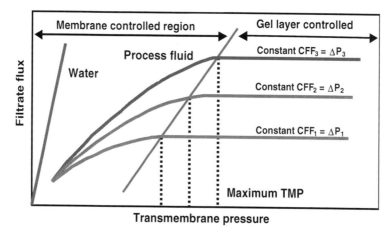

Figure 6.7. Effect of cross-flow rate on permeate flux.

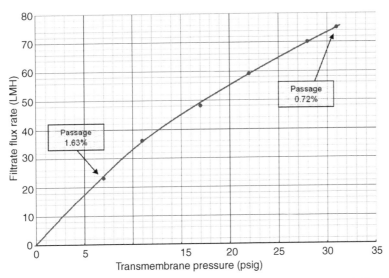

Figure 6.8. Effect of TMP on % transmission for r-hGH. Initial concentration 0.83 mg/mL on 5 kDa membrane.

pressure/membrane controlled region and delaying the onset of the gel layer controlled region to a higher TMP. Cross-flow turbulence can be used to characterize the TFF system and can be expressed in terms of cross-flow rate, shear rate, or Reynolds number.

Assuming that product retention is acceptable, the optimal operating point is in the transition region just prior to the gel layer controlled region. The transition point provides the optimal balance between flux performance and external pressure. The plot in Fig. 6.7 can be used to determine the transition region and the optimal TMP that maximizes permeate flux.

Higher cross-flow also results in greater product shear and heat generation and requires large pumps and piping. A suitable cross-flow flux is selected to give the desired permeate flux while balancing these other effects. To enable comparison with other membrane configurations and to aid in scale-up, the cross-flow rate is specified in terms of the cross-flow flux.

In addition, since TMP affects the gel layer, it is in effect a dynamic membrane. As TMP increases, the gel layer restricts solute transmission increasing the rejection coefficient. While this has a beneficial effect for product recovery as shown in Fig. 6.8, the retention of unwanted (contaminant) species also increases, which could result in longer diafiltration cycles to achieve the needed purity. Retention should again be assessed and the operating TMP adjusted as necessary to achieve the desired product retention and contaminant clearance.

Temperature influences fluid viscosity and density and solute diffusivity. As a result, increasing the operating temperature increases permeate flux. However, in biotechnology applications, temperature generally has detrimental effects on the product. It is thus important to assess the effects of temperature on the product and specify the highest possible temperature within product and membrane stability constraints.

6.6.1 QbD Operating Parameter Design Principles

1. Prepare Flux versus TMP curve for a given cross-flow rate.
   - Identify the transition region.
   - Calculate the mass transfer coefficient for the system.

$$J_{filtrate} = k \times TMP$$

   - Calculate product retention coefficient and verify contaminant clearance at high TMP.
   - Select the maximum TMP within the transition region.
2. Prepare flux versus TMP curves at increasing cross-flow rates.
   - Select cross-flow (CFF) that maximizes permeate flux within system pumping, piping, and cooling constraints.

$$CFF = \frac{Q_R}{A}$$

   where Q_R = retentate flow rate (L/min) and A = membrane module area (m^2).
3. Evaluate effect of temperature on product yield and stability. Select highest operating temperature to maximize filtrate flux.

6.7 TFF DIAFILTRATION OPERATING MODE DESIGN

Diafiltration is most commonly employed to transport contaminants from the retentate to the permeate, thereby removing them from the product. It can also be used to fractionate the product from larger or higher molecular weight species. Quality design of the diafiltration operating mode is critical to minimizing operating time, buffer volume, and for shear sensitive products, product recovery. Continuous diafiltration involves feeding diafiltration buffer to the retentate, while maintaining a constant retentate volume. Continuous diafiltration is more efficient than discontinuous diafiltration, and current equipment design and control have made discontinuous diafiltration largely obsolete.

To specify the number of continuous diafiltration volumes (diavolumes) required to reduce contaminants to acceptable levels, the rejection coefficient for the contaminant(s) should be determined at the selected operating parameters [22]. The contaminant rejection coefficient must be determined at this stage since it somewhat depends on the gel layer that is impacted by the transmembrane pressure and cross-flow flux. The number of diavolumes (with an over design factor of two diavolumes) can be calculated from

$$N = \frac{\ln\left(1 - \frac{Y_F}{100\%}\right)}{(R - 1)} + 2$$

where N = number of diavolumes, Y_F = % solute in permeate, and R = solute rejection coefficient.

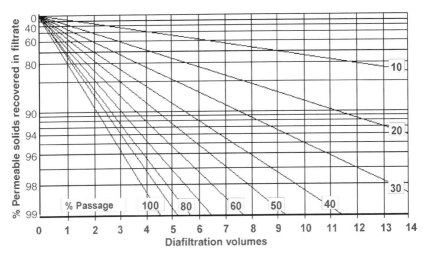

Figure 6.9. Solute transport to permeate stream as a function of passage (1 − rejection coeffi-cient) for continuous, constant volume diafiltration.

Figure 6.9 illustrates the transport of solute from the retentate to permeate as a function of the solute rejection. This represents ideal, theoretical mass transport. In actual operation, there exist system nonlinearities such as solute interactions and imperfect level and feed control that hinder solute transport. To ensure solute removal, engineering over design factor of two diavolumes is typically added to the calculated number of diavolumes [18]. Using such a factor represents good quality design by accounting for the system nonlinearities.

The optimal concentration at which to commence diafiltration is derived from the equation for permeate flux in the gel layer controlled region. From this equation, the optimal diafiltration concentration has been derived from the film theory model as [10]

$$C_D = \frac{C_G}{2.7}$$

where C_D = optimal retentate concentration for diafiltration and C_G = gel layer concentration, and C_G can be determined from a plot of permeate flux versus feed concentration as shown in Fig. 6.10. For proteins, C_G is generally in the range of 20–40% (200–400 g/L), which is approximately an order of magnitude higher than typical processing conditions. Operating at a feed concentration as close as possible to C_D will minimize the total diafiltration volume and processing time. Note that the slope of a plot of J versus $\ln(C_B)$ is the mass transfer coefficient, k, which can additionally be computed to further characterize the TFF system.

In practice, TFF systems are often operated by concentrating to the final desired concentration and performing diafiltration at this concentration. However, this may not be the most efficient operating mode since the flux at the final concentration may be quite low resulting in excessively long operating times. Generally, the most time efficient diafiltration operating mode is to concentrate the feed close to C_D, perform the required number of diafiltrations, and then concentrate or dilute to the final desired concentration.

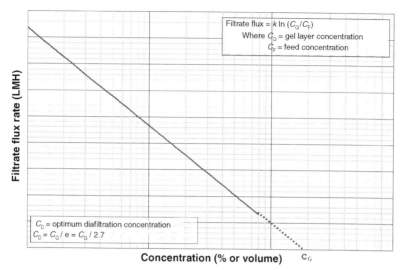

Figure 6.10. Determining the optimum concentration to perform diafiltration.

TABLE 6.1. Comparison of Different Concentration/Diafiltration Combinations for Processing Albumin. Objective 5× Concentration and 5× Diafiltration

	Batch Volume (L)	Protein Conc. (%)	Filtrate Rate (L/h)	Average Flow Rate (L/h)	Permeate Volume (L)	Filtrate Time (h)
CASE A						
Initial	1000	5	940			
Conc. 5×	200	25	140	460	800	1.7
Diaf. 5×	200	25	160	150	1000	6.7
Total					1800	8.4
CASE B						
Initial	1000	5	940			
Diaf. 5×	1000	5	1060	100	5000	5
Conc. 5×	200	25	160	520	800	1.5
Total					5800	6.5
CASE C						
Initial	1000	5	940			
Conc. 5×	500	10	600	740	500	0.7
Diaf. 5×	500	10	680	640	2500	3.9
Conc. 2.5×	200	25	160	370	300	0.8
Total					3300	5.4

Figure 2.1. Quality by Design is a systematic approach to development that begins with predefined objectives and emphasizes product and process understanding and process control based on sound science and quality risk management (ICH Q8R1). (See page 11 for text discussion of this figure.)

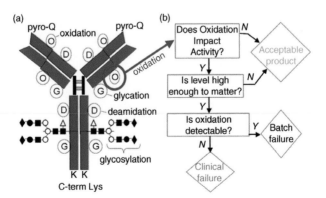

Figure 2.2. Biotechnology products have a large number of structural attributes that may impact product performance. (See page 13 for text discussion of this figure.)

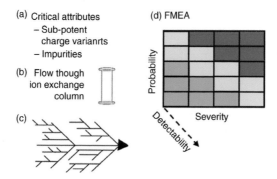

Figure 2.6. Generation of a design space starts with defining appropriate responses, the process to, and important process variables. (See page 19 for text discussion of this figure.)

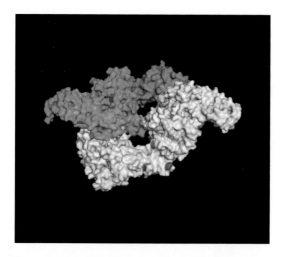

Figure 3.5. Space-filling representation of the dimeric structure of *P. falciparum* EBA-175 RII [47] showing positions of amino acid substitutions for expression of RII-NG. Monomers are shaded in cyanoblue and gray. (See page 39 for text discussion of this figure.)

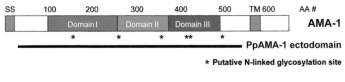

	Amino acid number				
	1 6 2	2 8 6	3 7 1	4 4 2 2 1 2	4 9 9
Native AMA-1	NTT	NYT	NAS	NNSS	NST
Syn P. pastoris AMA-1	KTT	NLV	NRE	NDKN	QST
Syn mammalian AMA-1	KTT	NYV	NAA	NQAS	QST

Figure 3.7. Scheme of gene structure of AMA1 showing division of domains I–III and other domains (upper panel). (See page 43 for text discussion of this figure.)

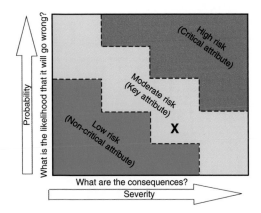

Figure 4.2. Example of criticality determination. (See page 56 for text discussion of this figure.)

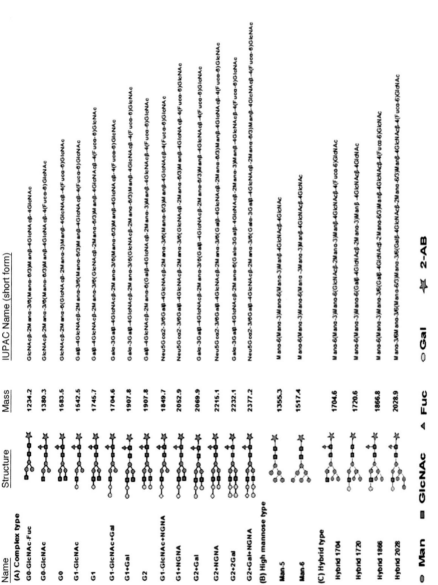

Figure 4.5. Oligosaccharide structures. (See page 60 for text discussion of this figure.)

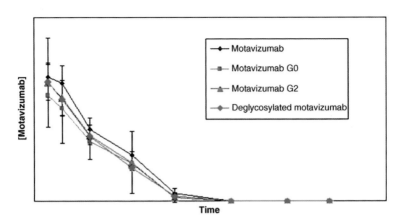

Figure 4.8. Cotton rat PK study: serum levels of motavizumab variants. (See page 64 for text discussion of this figure.)

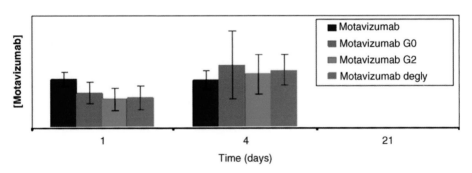

Degly = deglycosylated

Figure 4.9. Cotton rat PK study: BAL levels of motavizumab variants. Motavizumab is undetectable at 21 days. (See page 64 for text discussion of this figure.)

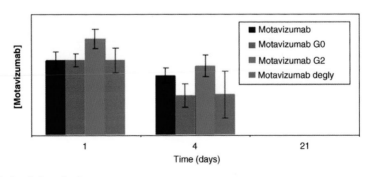

Degly = deglycosylated

Figure 4.10. Cotton rat PK study: lung homogenate levels of motavizumab variants. Motavizumab is undetectable at 21 days. (See page 64 for text discussion of this figure.)

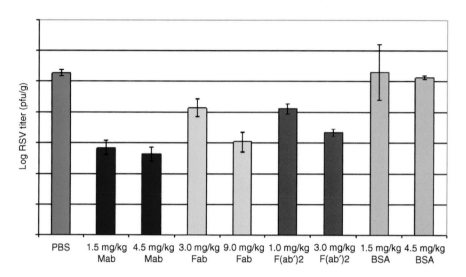

Figure 4.12. Intranasal prophylaxis of RSV infection in cotton rats by using palivizumab MAb, Fab, and F(ab')2 fragments. (See page 65 for text discussion of this figure.)

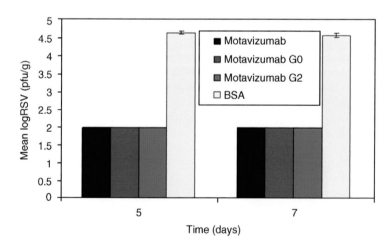

Lower limit of detection = 2 logs

Figure 4.13. IM prophylaxis in cotton rats by motavizumab-enriched glycoforms. (See page 66 for text discussion of this figure.)

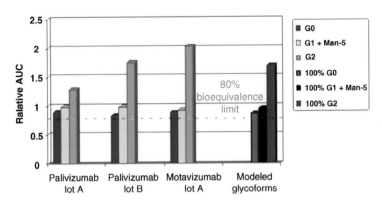

Figure 4.17 Modeling relative AUC for palivizumab and motavizumab glycoforms. (See page 69 for text discussion of this figure.)

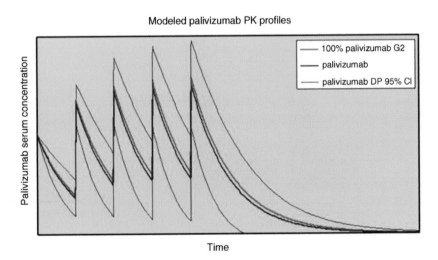

Figure 4.18. Seasonal modeling of palivizumab and G2 glycoform PK profiles. (See page 70 for text discussion of this figure.)

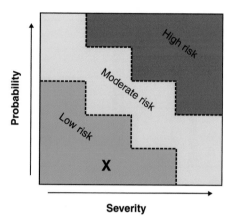

Figure 4.19. Criticality of glycosylation. (See page 72 for text discussion of this figure.)

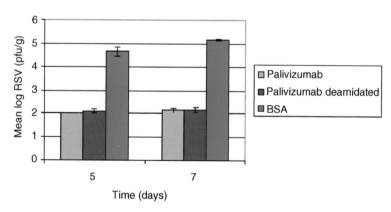

Figure 4.29. Prophylaxis of RSV infection in cotton rats by using palivizumab and buffer-deamidated palivizumab. (See page 78 for text discussion of this figure.)

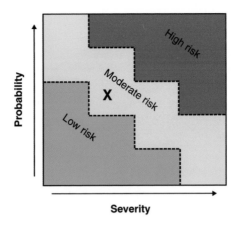

Figure 4.30. Criticality of deamidation. (See page 81 for text discussion of this figure.)

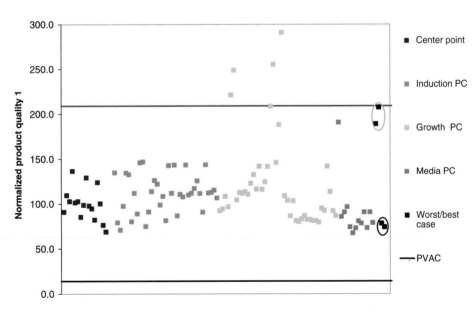

Figure 5.16. Projection of bench scale characterization data for product quality 1 onto the PVAC derived from Monte Carlo simulation. (See page 107 for color representation of this figure.)

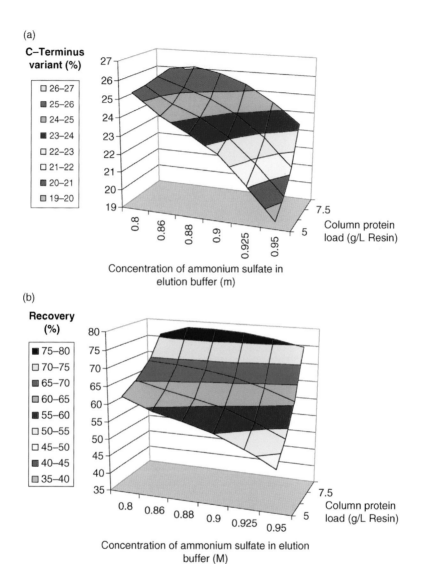

Figure 7.4. Hydrophobic interaction chromatography: the effect of column loading and ammonium sulfate concentration in the elution buffer on (a) the level of CTV in the column eluate and (b) product recovery. (See page 139 for text discussion of this figure.)

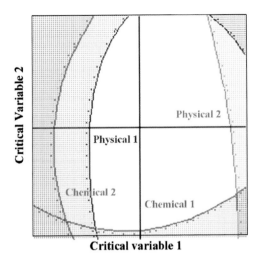

Figure 9.6. Concept of formulation design space. (See page 171 for text discussion of this figure.)

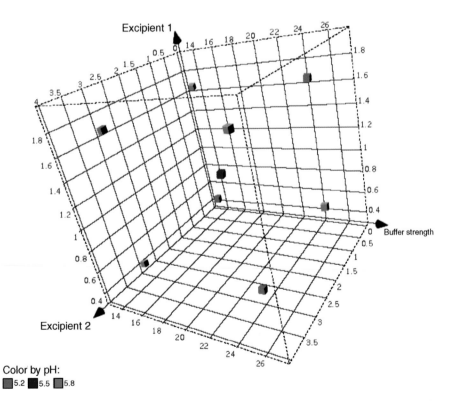

Color by pH:
■ 5.2 ■ 5.5 ■ 5.8

Figure 10.5. Experimental design for robustness of product composition. The center point represents the optimal or target composition. (See page 185 for text discussion of this figure.)

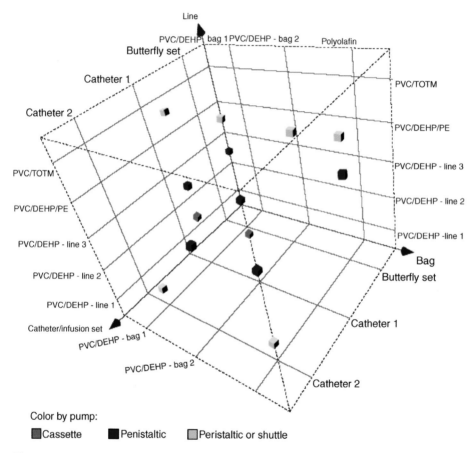

Figure 10.8. Experimental design for testing compatibility of infusion components with a biologic infusion product. (See page 191 for text discussion of this figure.)

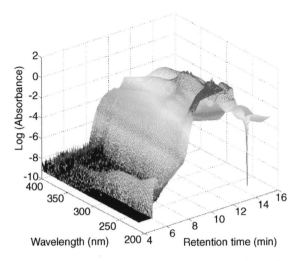

Figure 11.2. Example of a diode-array chromatogram showing the data cube generated for one injection of sample or control. (See page 201 for text discussion of this figure.)

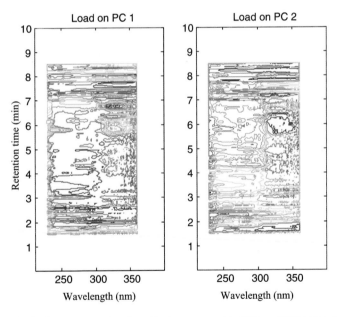

Figure 11.4. Principal component loadings that correspond to PC 1 and PC 2 in Fig. 11.3. PC 2 has an intensity pattern that suggests media additive Lot 1 (from Fig. 11.3) has more material absorbing at wavelengths greater than 300 nm and less material from 7.5 to 8.5 min. (Blue: less intense; red: more intense.) Not many differences are observed in the PC 1 loadings that reflect the sample scores plot. (See page 202 for text discussion of this figure.)

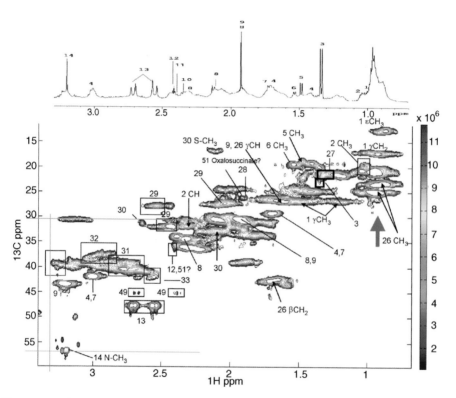

Figure 11.7. ^{1}H NMR (top) and HQSC NMR spectra (bottom) for a representative hydrolysate sample. Numbers indicate distinct functional groups or compounds that account for the signal. (See page 207 for text discussion of this figure.)

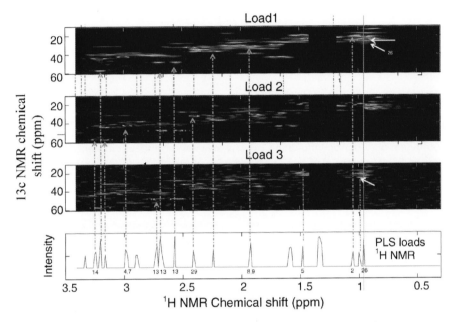

Figure 11.8. Comparison of three loadings from multiway PCA of HSQC with ¹H NMR loadings from a PLS analysis. Peaks that contribute significantly to the PLS analysis of ¹H NMR can be identified in the HSQC NMR loads. Identified compounds: 2, valine; 4, cadaverine; 5, free alanine; 7, lysine; 8, N-acetyl glutamic acid; 9, arginine; 13, citrate; 14, choline; 26, leucine; 29, acetyl phosphate. (See page 207 for text discussion of this figure.)

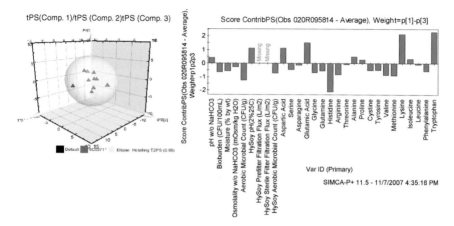

Figure 12.8. PC score 3D plot (top) showing various grades of raw materials and variable contribution plot (bottom) indicating the raw material constituent differences between the historical lots (black triangles) and a new lot (red triangle). (See page 235 for text discussion of this figure.)

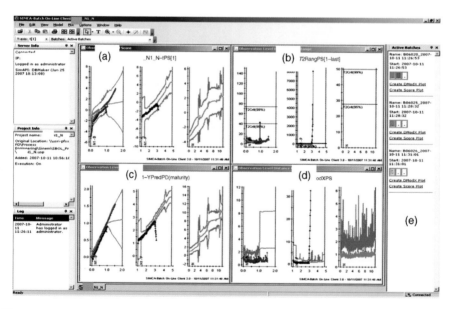

Figure 12.9. Online real-time detection and diagnosis of a process deviation enables engineers to troubleshoot the process more effectively. Panel (a) score time series plots, (b) T^2 time series plots, (c) batch maturity plots, (d) DModX (or squared prediction error) plots, and (e) executive batch dashboard for two consecutive bioreactors (first two plots or boxes in the dashboards for two phases in the first bioreactor, batch and fed-batch). (See page 238 for text discussion of this figure.)

This operating mode is illustrated by the case study shown in Table 6.1, where the operating time was reduced from 8.4 to 5.4 h. This operating mode will result in additional buffer usage for diafiltration, and hence the cost benefit of time versus buffer volume will ultimately determine the diafiltration operating mode.

6.7.1 QbD Principles for Operating Mode Design

Calculate number of diavolumes based on rejection coefficient.

$$N = \frac{ln\left(1 - \frac{Y_F}{100\%}\right)}{(R-1)} + 2$$

Prepare a plot of permeate flux versus feed concentration to determine the optimal diafiltration concentration.

$$C_D = \frac{C_G}{2.7}$$

Tabulate operating time and diafiltration buffer volumes as a function of diafiltration feed concentration. Select operating mode based on minimum operating time and diafiltration volume.

6.8 SUMMARY

Applying theoretical principles and mechanistic modeling can be used to establish a "quality" design for TFF operations. The proven design models for TFF can be used to identify optimal operating conditions as well as a solid understanding of the system and how the major operating parameters interact. The use of such mechanistic equations in a planned stepwise fashion will enable the user to optimize and understand the TFF system, thus achieving the goals of a Quality by Design approach.

REFERENCES

[1] Food and Drug Administration. "Pharmaceutical cGMPs for the 21st Century—A Risk Based Approach, Final Report, 2004. http://www.fda.gov/Cder/gmp/gmp2004/CGMP%20report%20final04.pdf.

[2] Food and Drug Administration. Innovation and Continuous Improvement in Pharmaceutical Manufacturing. Pharmaceutical CGMPs for the 21st Century. The PAT Team and Manufacturing Science Working Group Report, 2004. http://www.fda.gov/Cder/gmp/gmp2004/manufSciWP.pdf.

[3] Food and Drug Administration. Guidance for Industry PAT—A Framework for Innovative Pharmaceutical Development, Manufacturing, and Quality Assurance, 2004. http://www.fda.gov/cder/guidance/6419fnl.pdf.

[4] ICH Guidance for Industry: Q9. Quality Risk Management, 2005. http://www.ich.org/LOB/media/MEDIA1957.pdf.

[5] ICH Guidance for Industry: Q8. Pharmaceutical Development, 2006. www.fda.gov/cber/gdlns/ichq8pharm.htm.

[6] ICH Guidance for Industry: Q8 Annex. Pharmaceutical Development, Annex to Q8, 2007. http://www.ich.org/LOB/media/MEDIA4349.pdf.

[7] ASTM E 2500 – 7. Standard Guide for Specification, Design, and Verification of Pharmaceutical and Biopharmaceutical Manufacturing Systems and Equipment, 2007. http://www.astm.org/Standards/E2500.htm.

[8] Hall G, Runas R. JPI Interviews Moheb Nasr, PhD. *J Pharm Innov* 2007;2:67–70. http://www.springerlink.com/content/d65042r0h3k41641/fulltext.pdf.

[9] Hutchins R. The Great Conversation. The Substance of a Liberal Education. *Encyclopedia Britannica, Inc.*; 1952.

[10] Cheryan M. *Ultrafiltration and Microfiltration Handbook*. New York: CRC Press; 1998.

[11] Treybal RE. *Mass Transfer Operations*. New York: McGraw-Hill; 1980.

[12] Wang A, Lewus R, Rathore A. Comparison of different options for harvest of a therapeutic protein product from high cell density yeast fermentation broth. *Biotechnol Bioeng* 2006; 94: 91–104.

[13] Hwang S-T, Kammermeyer K. *Membranes in Separations*. New York: John Wiley & Sons; 1975.

[14] Gaddis JL, Jernigan DA, Spencer HG. Determination of gel volume deposited on ultrafiltration membranes. In: Sourirajan S, Matsuura T, editors. *Reverse Osmosis and Ultrafiltration*. Washington: ACS; 1985. p 413–427.

[15] Lydersen BK, D'Elia NA, Nelson KL. *Bioprocess Engineering*. New York: John Wiley & Sons; 1994. p 118–157.

[16] Ulber R, Plate K, Reif O-W, Melzner D. Membranes for protein isolation and purification. In: Hatti-Kaul R, Mattiasson B, editors. *Isolation and Purification of Proteins*. New York: Marcel Dekker; 2003. p 191–223.

[17] Ghosh R. *Protein Bioseparation Using Ultrafiltration: Theory, Applications and New Developments*. London: Imperial College Press; 2003.

[18] Millipore Corp. Protein Concentration and Diafiltration by Tangential Flow Filtration—An Overview. Technical Brief TB032. 2003. Millipore Corporation, Billerica.

[19] Rudolf EA, MacDonald JH. Tangential flow filtration systems for clarification and concentration. In: Lydersen K, D'Elia NA, Nelson KL, editors. *Bioprocess Engineering*. New York: John Wiley & Sons; 1994.

[20] Wheelwright SM. *Protein Purification—Design and Scale up of Downstream Processing*. New York: John Wiley & Sons; 1991.

[21] Ladisch MR. *Bioseparations Engineering—Principles, Practice and Economics*. New York: John Wiley & Sons; 2001.

[22] Rathore A, Sharma A, Chilin D. Applying Process Analytical Technology to Biotech Unit Operations. *BioPharm Int* 2006;8:48–57.

7

APPLICATIONS OF DESIGN SPACE FOR BIOPHARMACEUTICAL PURIFICATION PROCESSES

Douglas J. Cecchini

7.1 INTRODUCTION

The goal of biopharmaceutical process development is a robust manufacturing process that delivers consistent productivity and product quality. The evolution of process understanding, however, does not end at the time of biological license application and approval. Process understanding continues to evolve as manufacturing history grows, particularly with the emergence of more sophisticated tools and approaches for statistically based multivariate process monitoring [1] and process analytical technologies (PAT) [2, 3]. In the course of process monitoring, process limitations identified during scale-up or opportunities to improve process consistency and productivity are likely to be identified. The ICH Q8 guidance document [4] provides

> ...an opportunity to demonstrate a higher degree of understanding of material attributes, manufacturing processes and their controls. This scientific understanding facilitates establishment of an expanded design space.

Design space in turn is defined as

> ...the multidimensional combination and interaction of input variables (e.g., material attributes) and process parameters that have been demonstrated to provide

Quality by Design for Biopharmaceuticals, Edited by A. S. Rathore and R. Mhatre
Copyright © 2009 John Wiley & Sons, Inc.

assurance of quality. Working within the design space is not considered as a change. Movement out of the design space is considered to be a change and would normally initiate a regulatory post-approval change process. Design space is proposed by the applicant and is subject to regulatory assessment and approval.

Thus, an approved, expanded manufacturing design space will allow manufacturers of biopharmaceutical products the flexibility to take advantage of the high degree of process knowledge and implement post-approval process improvements with a reduced burden of regulatory submission and approval.

An expanded design space for biopharmaceutical purification processes (downstream processing) provides a number of advantages. In general, the increased process understanding that design space requires supports a fundamental premise of the quality by design initiative, namely "that quality cannot be tested into products, that is, quality should be built in by design" [4]. In other words, an enhanced understanding of the relationship between operational parameters and product attributes enhances quality assurance. Moreover, understanding the correlations between process parameters and product quality lessens the reliance on final product testing and provides the foundation for the advent of parametric release [5] for biopharmaceutical products.

In addition, movement within the purification design space could be utilized to better ensure product and process consistency. For example, changes to cell culture processes toward improved productivity or variation in cell culture performance due to raw material variability could alter product quality attributes. For downstream processing, lot-to-lot variability in chromatography resins or column-packing limitations at large scales could result in unanticipated changes in yield or product quality. An expanded design space based on an in-depth understanding of how operating parameters affect the process and product attributes could be used to ensure process consistency and/or that product quality attributes remain within predetermined specifications. Alternatively, design space could be used to accommodate operational flexibility associated with changing manufacturing facility requirements, economics, and schedules. For example, variable column size (height and diameter) and number of chromatography cycles per batch and the sizes and configurations of filters could all be incorporated into design space. In addition, the optional use of stable storage forms for process intermediates could be used to decouple manufacturing steps and provide flexibility in manufacturing schedules.

7.2 ESTABLISHING DESIGN SPACE FOR PURIFICATION PROCESSES DURING PROCESS DEVELOPMENT

A recommended sequence of activities toward the establishment of an expanded design space is outlined in Table 7.1. Most commonly, design space considerations will be applied to late-stage processes, that is, processes developed and optimized with the intention of validation in support of a Biological License Application and subsequent commercial production. As such, a certain amount of process knowledge and, in some cases, manufacturing experience is available at the outset of more in-depth process characterization. As shown in Table 7.1, the acceptable variability in critical product quality attributes must be established first. This assessment would consider, for example,

TABLE 7.1. Recommended Sequence of Activities for the Development of Design Space

Stage	Activity	Outcome
Determine acceptable ranges for product quality attributes	Identify critical product attributes and acceptable variation	Product attribute responses to be measured during process characterization (e.g., levels of CHO host-cell proteins, levels, aggregate levels, potency)
Risk analysis	Evaluate process risks with respect to product and process consistency and operational flexibility	Subset of unit operations to be explored for expanded design space
Parameter screening	Use existing process knowledge and previous experience with similar processes to eliminate nonkey parameters. Resolution III or IV screening DOE	Identification of key and critical operational parameters for modeling DOE.
Modeling DOE	Resolution V or response surface DOE to detect curvature (nonlinear responses)	Identification of interactions between variables Predictive small-scale model Failure limits Proposed limits for design space
Scale-down model verification	Comparison of modeling DOE predictions to manufacturing scale operation Satellite runs	Verified relevance of small-scale models to manufacturing scale process.

previous clinical exposure of the product, nonclinical (animal) studies, *in vitro* biological activity assays, an understanding of the molecule's mechanism of action, and manufacturing capability. Ultimately, it is the relationship between the variation in operational parameters (i.e., process inputs) and the variation in critical quality attributes, which determines the boundaries of the process design space. Next, process characterization (robustness) studies can be used to explore the operating ranges and establish acceptable ranges for the operational parameters. Once established, operating within these limits, the combination of which defines the design space, provides the "assurance of quality." It should be mentioned that for products produced in mammalian cell culture, the assurance of quality must include the demonstration of adequate viral clearance. Thus, the evaluation of process changes within the design space must include an assessment of how these changes affect the level of viral clearance. In cases where the factors controlling viral clearance for a particular step are not well understood, additional viral clearance studies would be required to justify the proposed design space.

The terms *space* and *multidimensional combination* in the ICHQ8 definition of design space imply the need for extensive use of design of experiments (DOE) to map primary effects and interactions between variables during process characterization studies. Moreover, the large numbers of variables that require testing are best managed through the simultaneous testing of multiple variables, which DOE provides. These experiments should cover wide ranges for process variables and material attributes and be designed to determine failure limits and acceptable ranges for the manufacturing process. An in-depth discussion of DOE can be found elsewhere [6, 7].

Given the number of variables associated with downstream processing of biopharmaceutical products, establishing an expanded design space for all operational parameters for all unit operations can be a daunting and ultimately unnecessary task. A risk assessment will help identify the subset of process steps most likely to benefit from an expanded design space. The selection criteria for such an analysis would include considerations such as (1) impact of a purification step on critical product attributes, (2) potential factors that impact process consistency (e.g., yield), and (3) opportunities for manufacturing operational flexibility. In line with standard risk assessment practices [8], these elements would be evaluated in terms of probability of occurrence and severity of the consequences and prioritized appropriately. The timing of such an analysis would depend on the availability of sufficient process and product understanding, but would best be performed with an established purification process prior to extensive process characterization.

Once process steps are selected, the early identification and focus on critical and key operational parameters for each process step can further reduce the workload to a manageable number of parameters that can be investigated with a reasonable amount of time and resources. "Key" and "critical" operating parameters are terms adopted from PDA Technical Report 42 [9]. Critical operating parameters affect critical product quality attributes when varied outside of a narrow (or difficult to control) operating range. Key operational parameters also have a narrow (or difficult to control) range. Key operating parameters, however, affect process performance (e.g., yield, duration), but not product quality. The remaining (nonkey) parameters can affect process or product but are easily controlled within wide acceptable limits. Much of this knowledge may be available from earlier process development and from previous experience with similar processes. At this point, however, if there is large number of process variables to consider, additional parameter screening with low-resolution DOE (resolution III or IV) [6, 7] may be required to identify the main effects. Parameters showing minor effects on process or product across wide ranges (i.e., nonkey parameters) are less pertinent to process control and can be excluded from further process characterization studies.

Screening studies are followed by higher resolution designs (e.g., resolution V or response surface) powered to address interactions between variables. In addition, if nonlinear responses to input parameters were detected during screening studies, response surface DOE should be used to map curvature of the operating space [6]. The goal of these higher resolution studies is to develop a predictive model, determine failure points and establish design space limits for critical and key parameters.

Given the number of experiments required to establish design space, the majority of these studies will take place at laboratory scale. Thus, the successful implementation of the design space aspect of quality by design is contingent upon carefully designed scale-down models that are representative of the manufacturing-scale operation. The scale-down models should be designed such that, to the extent possible, key and critical parameters are maintained across scales. Moreover, the performance of the laboratory-scale models used to develop the design space needs to be representative of the manufacturing-scale process with regard to process performance and product comparability [10–12]. A comparison of the manufacturing-scale process data to the predictive models developed during process characterization will provide evidence for the validity of the scale-down models. When manufacturing data do not exist, this comparison will need to be done retrospectively after the process has been performed at full scale. If sufficient manufacturing-scale data exist, predetermined acceptance criteria can be developed prior to the scale-down runs or it may be demonstrated that the small-scale data are within the full-scale historical range [11]. Alternatively, for a true side-by-side comparison, the validity of models can be demonstrated with a scaled-down "satellite" process that is run in parallel with and starting with cell culture media taken from the full-scale manufacturing process.

7.3 APPLICATIONS OF DESIGN SPACE

Several potential uses of design space are illustrated in the context of a typical manufacturing process shown in Fig. 7.1 for a monoclonal antibody (mAb) or an antibody-like (fc) fusion protein (FP) produced in mammalian cell culture. The process consists of cell removal by filtration or continuous flow centrifugation, followed by a recombinant protein A affinity "capture" column, then a low pH hold for virus inactivation, and subsequently further purification by one or two additional chromatography steps. The latter two steps commonly comprise some combination of anion exchange, cation exchange, hydrophobic interaction, or ceramic hydroxyapatite chromatography. Following the chromatography steps, the product is passed through a virus-removal filter and subsequently concentrated and transferred (by diafiltration) into the formulation buffer by using cross-flow ultrafiltration (UF). The purification process provides the removal of process and host-cell related impurities such as cell culture additives, host-cell protein, and DNA. In addition, the purification process may control the level of product-related impurities and isoforms including clipped species, charge variants, aggregates, and glycoforms. Lastly, for proteins produced in mammalian cell culture, the downstream process must provide assurance that retroviruses and other adventitious agents potentially present during cell culture are sufficiently removed by the purification process [13, 14]. In the remainder of this chapter, several examples of how design space could be applied to downstream processing are described. Each example is taken from an actual manufacturing process for an antibody or fusion protein similar to that outlined in Fig. 7.1.

7.4 CELL HARVEST AND PRODUCT CAPTURE STEPS

Similar to the process outlined in Fig. 7.1, the production process for a humanized monoclonal antibody (mAb1) uses continuous-flow centrifugation for the cell removal

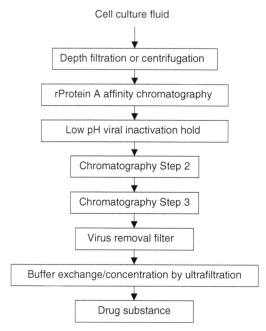

Figure 7.1. Typical process flow for the purification of a mAb or fusion protein. Chromatography columns 2 and 3 commonly comprise some combination of anion exchange, cation exchange, hydrophobic interaction, or ceramic hydroxyapatite chromatography.

(harvest) step, followed by mAb1 capture on a recombinant protein A affinity column (rPA) column. The rPA column is loaded with clarified cell culture conditioned medium treated sequentially with two intermediate wash buffers, followed by product elution at low pH. The manufacturing process for mAb1, which was used to support late-stage (pivotal) clinical studies, used an intermediate wash step on the rPA column containing a high-concentration chaotrope to enhance the clearance of CHO host-cell protein (CHOP). Risk analysis associated with a change of manufacturing facility and manufacturing scale revealed that this reagent was expensive, difficult to handle, and an environmentally unfavorable component of the waste stream. Consequently, an expanded design space was investigated with the goal of reducing or eliminating the need for the use of chaotrope in the intermediate wash steps for the rPA column.

Preliminary studies had shown that adjustment of the cell culture harvest medium to low pH causes cells and cellular debris to flocculate and the resulting particulates are more efficiently removed by centrifugation [15, 16]. This reduces the turbidity of the centrate and minimizes filtration area requirements prior to rPA chromatography. In addition, the pH-induced flocculation results in clearance of CHOP and host-cell DNA, and the levels of these impurities are significantly reduced in the centrate produced at low pH. These findings are illustrated in Fig. 7.2. Harvesting the cell culture fluid at pH 5.0 or lower reduced DNA levels in the centrate to below the level of detection, while the levels of CHOP were reduced two- to fivefold. Harvest pH did not, however, significantly alter

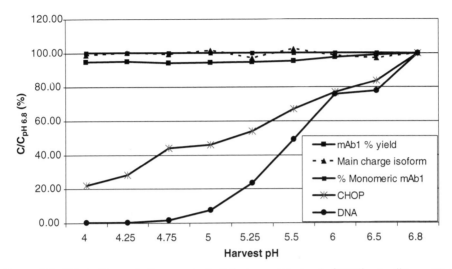

Figure 7.2. Effect of harvest pH on product attributes and the level of CHO host-cell impurities after centrifugation. The *y*-axis represents the levels of product and impurities in the centrate, normalized to the pH 6.8 harvest condition. Adjustment of the cell culture harvest medium to acidic pH causes cells to flocculate, resulting in more efficient centrifugation. In addition, the pH-induced flocculation results in reduced levels of CHO host-cell protein (CHOP) and DNA in the centrate. Harvest at reduced pH did not alter the product (mAb1) recovery, charge profile, or level of aggregation.

the product concentration in the centrate or product quality attributes such as aggregation and charge variants.

These findings indicate that harvest pH affects the purity of the load onto the rPA capture column and the harvest–capture design space was explored together in a series of DOEs. Other centrifugation parameters (e.g., g-force and flow rate) did not significantly alter the properties of the filtered centrate (rPA column load) across relatively wide ranges of operation. These parameters were not a significant source of process variation and offered little or no opportunity for process control. Thus, these "nonkey" parameters were not included in the subsequent design space studies. As mentioned earlier, product recovery and product attributes are relatively insensitive to the harvest pH. In addition, DNA is easily removed by the rPA and subsequent chromatography steps. Thus, the primary response for the DOE studies was determined to be the clearance of CHOP. The sequence of these experiments is summarized in Table 7.2. Briefly, of the 19 input parameters identified for the operation of the rPA capture column, 7 were identified as nonkey parameters based on prior process knowledge and were not carried forward into the resolution IV screening design. Of the 12 parameters tested in the screening DOE, 4 were shown to have significant effects on the process or the product and carried forward into the subsequent modeling design. The experimental design matrix for the modeling DOE is shown in Table 7.3. For this exercise with four input variables, a D-optimal design comprising 32 chromatographic runs carried out in two blocks was selected. This orthogonal response surface design (power >80 and 95% confidence interval) was

TABLE 7.2. Experimental Sequence for the Harvest / Capture Design Space

Process Step	Factor		Factor		Factor
Equilibration	velocity NacL conc. pH Volume				
Load Clarified Conditioned Media	Harvest pH Load pH capacity velocity		Harvest pH Load pH capacity velocity		Harvest pH capacity
Wash I	pH Urea conc NaCL Volume velocity		Urea conc NaCL Volume		Urea conc
Wash II	pH Velocity Volume		pH		
Elution	pH Velocity Volume		Volume pH Velocity Volume		Velocity

Eliminate factors based on existing process knowledge

Carry remaining 12 factors into resolution IV screening design (36 runs)

Eliminate factors based on screening design results.

Carry remaining 4 factors into response-surface modeling design (32 runs)

TABLE 7.3. Experimental Matrix for the Harvest–Capture Modeling DOE

Standard Order	Block	Harvest pH	Chaotrope Concentration	Elution Velocity	Load Capacity
1	1	0	0	0	0
2	2	− 1	− 1	− 1	1
3	1	1	− 1	1	0.
4	2	1	− 1	− 1	− 1
5	2	− 1	− 1	0	− 1
6	2	− 1	1	− 1	− 1
7	2	− 1	1	1	0
8	1	1	1	0	− 1
9	2	1	1	− 1	1
10	2	− 1	1	0	1
11	2	1	1	1	1
12	1	− 1	− 1	1	1
13	1	− 1.	0	1	− 1
14	1	1.	− 1	0	1
15	1	0	1	1	− 1
16	2	0	0	0	0
17	1	0	0	0	0
18	2	0	0	0	0
19	1	0	0	0	0
20	2	0	0	0	0
21	1	0	− 1	1	− 1
22	1	1	0	1	− 1
23	2	− 1	− 1	− 1	− 1
24	1	1	1	− 1	− 1
25	1	− 1	− 1	1	− 1
26	1	1	− 1	1	− 1
27	2	1	− 1	0	1
28	2	− 1	1	− 1	− 1
29	1	1	1	1	1
30	2	− 1	− 1	1	1
31	1	− 1	− 1	− 1	1
32	2	1	− 1	− 1	− 1.

Note: This D-optimal response surface design [7] comprises 32 chromatographic runs carried out in two blocks (power > 80 and 95% confidence interval). The −1, 0, and 1 designations refer to the low, center-point, and high end of the range, respectively, for each of the four input variables tested.

selected to detect nonlinear effects, secondary interactions between variables, and allow for independent estimation of the effect of each input variable on the response [7]. Two of the input parameters tested, harvest pH and chaotrope concentration during the first intermediate column wash, have a significant influence on CHOP levels in the capture column eluate. The remaining two parameters, column load and elution linear velocity, affect the column elution volume.

The outcome of these experiments indicates the optional use of a high-concentration (4 M) chaotrope intermediate wash step during the operation of the rPA capture column.

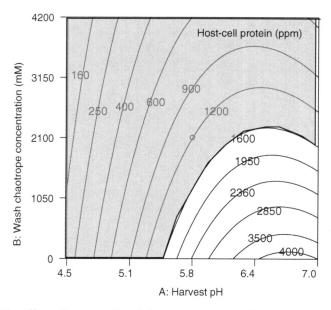

Figure 7.3. The effect of harvest pH and chaotrope concentration in the rProtein A capture column intermediate wash buffer on CHO host-cell protein (CHOP). The iso-lines represent constant levels of host-cell protein in the rProtein A column eluate. The grayed area represents the design space. Working within this region results in concentrations of CHOP that can be reduced to acceptable levels in the drug substance by the subsequent purification steps.

The relationship between harvest pH and the concentration of chaotrope required for the rPA wash step is illustrated in Fig. 7.3. Decreasing the harvest pH from 7.0 to 5.8 or below reduces the level of CHOP in the rPA column eluate from approximately 3500 ppm (ng/mg protein) to 1600 ppm or less, even in the absence of the chaotrope wash. At this reduced level of CHOP in the rPA eluate, the subsequent chromatography steps will provide sufficient reduction of CHOP to achieve the required levels of residual CHOP in the purified drug substance.

The use of the chaotrope-free process is clearly the preferred manufacturing option. The decision was made, however, to proceed to process validation with the original process and include the improved process as an alternate harvest–capture protocol in the initial biological license application. Obtaining regulatory approval of the alternate harvest–capture protocol prior to implementation will minimize the regulatory risks associated with making process changes late in clinical development.

7.5   PROTEIN A CAPTURE COLUMN

Protein A resins exhibit relatively low binding capacity (25–40 g product/L of resin) and thus require a large column or a smaller column with multiple chromatography cycles to capture the entire amount of product produced in cell culture. This is of

TABLE 7.4. Impact of Column Diameter on Cost and Operation of the Protein A Column

Column Diameter (cm)[a]	Resin Bed Volume (L)	Approximate Resin Cost (million dollars)[b]	Number of Chromatography Cycles Required[c]	Eluate Pool Volume (L)	Total Processing Time (h)[d]
140	308	3.1	13	10006	20
160	402	4.0	10	10053	15
200	628	6.3	6	9425	9

[a]Twenty centimeters bed height.
[b]Assumes an approximate cost of $10,000 per liter of rProtein resin.
[c]Cell culture volume 15,000 L; titer 9 g/L; rPA resin binding capacity 36 g/L resin.
[d]Linear flow rate 250 cm/h; processing time does not include support activities such as filter changes, buffer preparation, and so on.

increasing concern, given that cell culture titers in excess of 5 g/L are becoming increasingly regular [17]. At large scales and with the highly productive cell culture processes, large columns are required to capture the entire quantity of product produced in cell culture and rPA column can cost several millions of dollars. Conversely, running multiple cycles on a smaller column increases the complexity and duration of this step and, if taken to the extreme, can result in a significant product throughput bottleneck.

To help mitigate these concerns, flexibility in the size of the rPA column would allow the cost and duration of this step to be tailored as the economics and throughput demands of a product evolve throughout the product life cycle. For example, mAb2 has been produced in cell culture at a titer of 9–10 g/L and exhibits a dynamic binding capacity on the rPA column of approximately 36 g/L. The proposed design space for this product would allow flexibility in the choice of column diameter and the number of chromatography cycles run per production batch. The relationship between column diameter, the number of chromatography cycles required, protein A resin cost, and processing time for mAb2 is shown in Table 7.4. To ensure that these changes do not affect the process or product, the critical and key parameters required to ensure consistent performance of the column have been identified and will be maintained regardless of the column diameter. For example, column-packing parameters, protein load, and mobile-phase composition and linear velocity would remain within acceptable limits regardless of column diameter. In addition, the stability of the product in the capture column load and eluate is adequate to allow for variation in the duration of this step associated with more or less column cycling. Lastly, the volume (Table 7.4) and composition of the rPA eluate pool will not significantly change, and therefore will not affect the performance of the subsequent chromatography step.

7.6 HYDROPHOBIC INTERACTION CHROMATOGRAPHY

This third example pertains to purification of a FP, which employs a hydrophobic interaction chromatography (HIC) step. Ammonium sulfate is added to the column load to give a final concentration of approximately 1.0 M and the product is loaded onto the HIC column. The column is subjected to an intermediate wash step containing 1.0 M ammonium sulfate, and the product is then recovered with an elution buffer containing a reduced ammonium sulfate concentration. The HIC step reduces the level of host-cell related impurities, aggregated FP, and the level of a slightly more hydrophobic C-terminus variant (CTV).

Relatively narrow release specifications were established for the level of the CTV allowed in the purified drug substance on the basis of previous clinical exposure and a limited amount of manufacturing experience. As additional manufacturing experience was gained, however, variability in the amount of the CTV produced during cell culture was observed. This resulted in variability in the amount of the CTV in the HIC column load and eluate, which resulted in the occasional failure of the release specification for the CTV in the drug substance.

The reduction of the CTV form of FP on the HIC column is primarily controlled by column loading and the concentration of ammonium sulfate in the elution buffer (Fig. 7.4a). As anticipated, increased ammonium sulfate concentration in the elution buffer increases the retention of the more hydrophobic CTV, which results in improved clearance. Loading the column to lower levels of protein also improves the removal of the CTV. These two variables interact such that the effect of ammonium sulfate concentration on clearance of CTV is more pronounced at a lower column loading. As expected, increased ammonium sulfate concentration in the elution buffer decreases the overall recovery of the product. This effect is dramatic at low column loadings and minimized by operating at the higher end of the column loading range (Fig. 7.4b). Consequently, the preferred design space would entail a flexible ammonium sulfate concentration in the elution buffer, with column loading maintained at the higher end of the range (>7.5 g protein/L resin). The opportunity to vary the ammonium sulfate concentration in the elution buffer from 0.75 to 0.95 M would allow the column to be operated under conditions that optimize the trade-off between CTV removal and step yield as needed. It should be mentioned that viral clearance has been shown to be comparable across the proposed range of elution buffer concentrations. Ideally, implementation of the HIC design space would be supported by process analytical technologies. Specifically, a rapid measurement of the level of the CTV in the bioreactor or after the rPA capture column could be used to prompt a decision as to the concentration of ammonium sulfate to be used in the HIC column elution buffer.

7.7   ANION EXCHANGE CHROMATOGRAPHY

This final example pertains to the purification of a humanized monoclonal antibody (mAb3) by anion-exchange (AEX) chromatography. The product is loaded onto the column and the column is subjected to an intermediate wash step with slightly elevated conductivity. The intermediate wash step was developed to remove low levels of more basic charged isoforms produced during cell culture. The basic isoforms, however, are not well resolved and even under fully optimal conditions some of the desired product (~4%) is removed along with the undesired charged isoforms. Further increase in losses

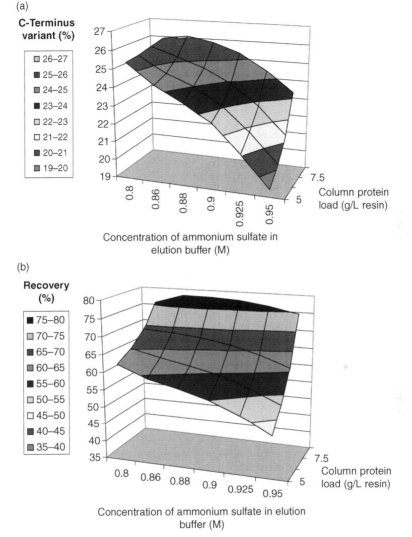

Figure 7.4. Hydrophobic interaction chromatography: the effect of column loading and ammonium sulfate concentration in the elution buffer on (a) the level of CTV in the column eluate and (b) product recovery. (See the insert for color representation of this figure.)

of protein during the wash step does not result in a measurable improvement in the clearance of the basic isoforms, but only serves to decrease product recovery. Following the intermediate wash step, the product is recovered from the column with an elution buffer of increased ionic strength.

During pilot-scale testing of the purification process, unanticipated variation in product recovery was observed. Step yields were up to 20% lower than that predicted by small-scale development studies. The yield loss was due to an atypically high loss of the product during the intermediate wash step. Further investigation indicated that the lot-to-lot variation in the AEX chromatography resin was the root cause of the yield variation. As shown in Fig. 7.5, the lot-to-lot variability in the ionic capacity of the resin is correlated to the loss in yield, with higher ionic capacity resin more prone to yield loss during the intermediate wash step. While the reason for this has not been investigated, ligand overcrowding and steric effects have been shown to reduce binding capacity on a number of adsorbent types [18].

The ionic capacity of each lot of AEX resin is reported on the vendor certificate of analysis and is predictive of resin performance (Fig. 7.5). The operating conditions for the AEX step can therefore be modified to modulate product losses during the wash step according to ionic capacity for the specific lot of resin to be used. The amount of protein removed during the intermediate wash step is dependent on the interaction of three factors—the pH and ionic strength of the wash buffer and the level of protein loaded onto the column. As one would predict, lower pH, and higher conductivity favor higher losses of product during the wash step and more protein is lost during the wash step at higher column loading. From a design space perspective, column loading was selected as the simplest means of compensating for the variability in resin ionic capacity. As shown in Fig. 7.5, the dependency of yield loss on ionic capacity during the wash step drops off dramatically at column loadings below 20 g protein/L resin. For resin lots with higher ionic capacity (e.g., >150 µeq/mL), column loading can be easily maintained below 20 g/L resin by increasing the number of chromatography cycles used to process all of the product from 2 to 3 or 4 cycles per batch. When using resins of low ionic capacity (e.g., <150 µeq/mL) the preference would be to process all of the material in two AEX cycles. The stability of the product in the AEX column load and eluate is adequate to allow for

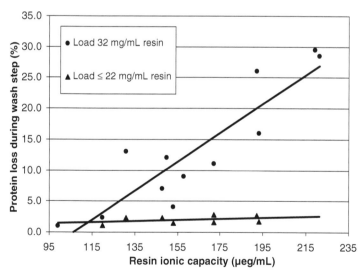

Figure 7.5. Anion exchange chromatography: lot-to-lot variability in resin ionic capacity can result in variable amounts of product loss during the intermediate wash step. At decreased column loading the effect of resin ionic capacity was negligible.

variation in the duration of this step associated with more or less column cycling. This flexibility in column loading and the number of chromatography cycles/batch would allow the trade-offs between step yield, buffer usage and processing time to be managed according to the lot of AEX resin that is in use. The potential effect of a variable load volume onto the subsequent chromatography step would need to be well understood before implementation of this aspect of the design space.

7.8 SUMMARY

An expanded, approved design space will allow for postapproval process changes (within the design space) with a reduced burden of regulatory submission. This added flexibility will allow manufacturers of biopharmaceutical products to better leverage process understanding gained during process development and subsequent manufacturing experience. Establishing design space during process development entails process characterization studies with verified scaled-down systems that accurately model the performance of the manufacturing-scale process. Design of Experiments is required to detect interactions between variables, establish predictive models and ultimately establish operating ranges for key and critical operational parameters. Risk assessment will help identify the subset of process steps most likely to benefit from an expanded design space and early identification of critical and key parameters for these process steps will further focus the experimental scope. The examples provided in this chapter demonstrate how design space can be applied to biopharmaceutical purification processes for improved control of product and process and to provide valuable flexibility to manufacturing operations.

ACKNOWLEDGMENTS

The author would like to thank members of the purification process development group at Biogen Idec, most notably Jorg Thommes, Lynn Conley, and John Pieracci, for their suggestions and contributions.

REFERENCES

[1] Sarolta A, Kinley RD. Multivariate statistical monitoring of batch processes: an industrial case study of fermentation supervision. *Trends Biotechnol* 2001;19:53–62.

[2] Ganguly J, Vogel G. Process analytical technology (PAT) and scalable automation for bioprocess control and monitoring—a case study. *Pharm Eng* 2006;26(1).

[3] U.S. Food and Drug Administration. Guidance for Industry: PAT-A Framework for Innovative Pharmaceutical Development, Manufacturing, and Quality Assurance. 2004. (http://www.fda.gov/CDER/GUIDANCE/6419fnl.htm)

[4] U.S. Food and Drug Administration. Guidance for Industry: Q8 Pharmaceutical Development. 2006. (http://www.fda.gov/CbER/gdlns/ichq8pharm.htm)

[5] The European Agency for the Evaluation of Medicinal Products, Committee for Proprietary Medicinal Products (CPMP): Note for Guidance on Parametric Release (CPMP/QWP/3015/99). 2001. (http://www.emea.europa.eu/pdfs/human/qwp/301599en.pdf)

[6] Kelly BD. Establishing process robustness using designed experiments. In: Sofer Gail, Zabriskie Dane W, editors. *Biopharmaceutical Process Validation*. New York: Marcel Dekker, Inc.; 2000. p 29–59.

[7] Anderson AJ, Whitcomb PJ. *DOE Simplified: Practical Tools for Effective Experimentation*. Portland, OR: Productivity Press; 2000.

[8] U.S. Food and Drug Administration. Guidance for Industry: Q9 Quality Risk Assessment. 2006. (http://www.fda.gov/cber/gdlns/ichq9risk.htm)

[9] Parenteral Drug Association. PDA Technical Report 42. Process validation of protein manufacturing. *PDA J Pharm Sci Technol* 2005;59(S-4):1–28.

[10] Gardner AR, Smith TM. Identification and establishment of operating ranges of critical variables. In: Sofer Gail, Zabriskie Dane W. editors. *Biopharmaceutical Process Validation*. New York: Marcel Dekker, Inc; 2000. p 61–76.

[11] Rathore AS, Krishnan R, Tozer S, Smiley D, Rausch S, Seely J. Scaling down of biopharmaceutical unit operations—part 2: chromatography and filtration. *Biopharm Int* 2005;18(4):58–64.

[12] Godavarti R, Petrone J, Robinson J, Wright R, Kelley BD. Scaled-down models for purification processes: approaches and applications. In: Rathore AS, Sofer G, editors. *Process Validation in Manufacturing of Biopharmaceuticals*. Boca Raton, FL: Taylor & Francis; 2005. p 69–142.

[13] Darling AJ. Validation of the purification process for viral clearance evaluation. In: Sofer Gail, Zabriskie Dane W, editors. *Biopharmaceutical Process Validation*. New York: Marcel Dekker, Inc; 2000. p 157–196.

[14] U.S. Food and Drug Administration. Guidance for Industry: Q5A Viral Safety Evaluation of Biotechnology Products Derived From Cell Lines of Human or Animal Origin. 1998. (http://www.gda/gov/cder/Guidance/Q5A-fnl.PDF)

[15] Romero, JK, Chrostowski, J, Vilmorin P. US PATENT (P0697 US 001—Prov. Appl. 60/855,734). Method for Isolating Biomacromolecules Using Low pH and Divalent Cations. U.S. Patent and Trademark Office, November 2, 2006.

[16] Romero, J, Chrostowski, J, Vilmorin, P Case, J. Effects of pH and ionic conditions on microfiltration of mammalian cells: combined permeate flux enhancement and mAb purification capabilities. American Institute of Chemical Engineering annual meeting, San Francisco, CA; 2006.

[17] Birch JR, Racher AJ. Antibody production. *Adv Drug Deliv Rev* 2006;58:671–685.

[18] Finette GMS, Mao QM, Hearn MTW. Comparative studies on the isothermal characteristics of proteins adsorbed under batch equilibrium conditions to ion-exchange, immobilised metal ion affinity and dye affinity matrices with different ionic strength and temperature conditions. *J Chromatogr A* 1997;763:71–90.

8

VIRAL CLEARANCE: A STRATEGY FOR QUALITY BY DESIGN AND THE DESIGN SPACE

Gail Sofer and Jeffrey Carter

8.1 INTRODUCTION

Although the concepts of design space and Quality by Design (QbD) go back at least to 1992, efforts to promote their formal adoption in biopharmaceuticals are more recent [1]. In fact, ICH Q8 was finalized in 2005 and the Annex to ICH Q8 is in draft form. The rate and extent to which the industry will ultimately adopt these principles are uncertain. The QbD and design space concepts relate broadly to product quality, which comprises "the suitability of either a drug substance or drug product for its intended use" [2]. Discussions relating QbD and design space concepts to virus clearance unit operations are sparse in the literature. This chapter aims to achieve two main purposes: explore benefits, limitations, and issues of applying ICH Q8 principles to virus clearance in biopharmaceutical manufacturing, and specifically to constructing a virus clearance process design space, and present a framework strategy for constructing a virus clearance design space.

8.2 CURRENT AND FUTURE APPROACHES TO VIRUS CLEARANCE CHARACTERIZATION

Manufacturers who derive guidance from the ICH Q5 guidance on viral safety currently take a multifaceted approach to ensure that they produce a product with an appropriate

Quality by Design for Biopharmaceuticals, Edited by A. S. Rathore and R. Mhatre
Copyright © 2009 John Wiley & Sons, Inc.

level of virus safety [3]. The complementary means of achieving this end point include the following:

- Identification of potential sources of viral contamination from, for example, cell banks or adventitious contamination during processing.
- Cell line qualification with respect to virus contamination.
- Testing of unprocessed bulk solutions and suspensions, for example, cell harvest pools and spent cell culture media.
- Assessment of specific unit operations within the manufacturing process for capability to effect virus clearance.

In this framework, it is common for a purely empirical understanding of process capability to be derived and therefore presented in a regulatory submission. Especially focusing on the assessment of virus clearance capabilities, it is typical for such characterization to be limited to a relatively small set of experiments (often three) in which scale-down models of a given unit operation are evaluated at nominal midpoint conditions or obvious worst-case settings of the clearance mechanism (e.g., lower temperature limit in a heat inactivation step or high pH limit in a low pH inactivation step). Data derived from these studies are used to support the claim that the process provides adequate virus clearance. The benefits of this approach are that regulators accept it and the costs to implement are relatively low.

The current approach is conceptually juxtaposed with a design space approach, a more holistic approach to characterizing the virus clearance capabilities of a unit operation and a process. A design space is conceptualized as a multidimensional window of operation that is sufficiently well characterized to allow processing anywhere within that window. The design space approach combines existing knowledge from the literature, one's technical understanding, and a stronger experimental design to achieve deeper knowledge of process capabilities. A brief comparison of current and design space approaches is found in Table 8.1. In this table, an intermediate milestone between the current state and the future end state is described to acknowledge that progress toward the end state should be viewed as a continuum and not as a step function.

8.3 BENEFITS OF APPLYING DESIGN SPACE PRINCIPLES TO VIRUS CLEARANCE

Although the benefits of QbD have been described by regulatory and industry representatives, the extent to which the industry and regulators will adopt QbD principles is not yet certain. The following brief discussion presumes that the adoption of QbD will be transformative in the industry.

The creation of a design space for a virus clearance unit operation requires that the manufacturer acquire in-depth knowledge of the unit operation technology in general and how that technology performs for the specific application. Regulators who agree that a

TABLE 8.1. Essential Differences Between Current and Design Space Process Characterization

	Technical Approach	Process Variability	Typical Learning
Current	Empirical	Not systematically assessed	If the manufacturing process conditions match the test conditions, virus clearance is reasonably assured
Milestone	Statistical and mechanistic	Main effectors understood	The impact to changes of the key independent variables is understood. Manufacturing deviations and process improvements may be justifiable based on knowledge generated
Design space (end state)	Statistical and mechanistic	Full system understood	As above, and with a full understanding of all effectors and interactions between independent variables

manufacturing process design space has been effectively established will have greater confidence that the manufacturer is making decisions from a strong knowledge base. This will lessen the requirement for regulatory oversight, allowing regulators to reallocate resources. In parallel, manufacturers gain freedom to change manufacturing processes within a multidimensional window without onerous regulatory consequences. The ultimate benefits of this better process understanding must translate directly to the patient through consistent viral safety assurance.

8.4 TECHNICAL LIMITATIONS RELATED TO ADOPTION OF QbD/DESIGN SPACE CONCEPTS IN VIRUS CLEARANCE

Developing an in-depth knowledge base for a virus clearance technology requires tools and technologies for virus clearance quantification. Several of these technologies have limitations that directly impact the cost and feasibility of achieving a fully characterized virus clearance design space.

8.4.1 Definition of Product CQA

A QbD approach requires one to set critical quality attributes (CQAs). In areas such as potency, purity, stability, and so on, the end points are assignable in real number, nonzero terms. The virus clearance level CQA is commonly stated as "zero virus in the product" or a "sufficient level of virus safety." Neither of these end points is actionable. The former suffers from the statistical issues inherent in claiming "zero" and the latter lacks quantification. This issue is faced by the industry today, and finds strong analogy in (e.g., steam) sterilization processes, for which the industry has established standards for quantifying process effectiveness.

8.4.2   Unknown Starting Virus Load

Once an acceptable virus clearance level CQA is targeted, the manufacturing process is evaluated for its capacity to clear viruses. The targeted virus clearance level is compared to the virus clearance capability to determine if the process provides an adequate safety margin. Such final determination requires a reasonable estimate of the starting virus contaminant load. For endogenous viruses, it is possible to make a reasonable estimate of starting virus load, as these viruses should be routinely expressed in each fermentation run. The handling of adventitious viruses is much more subjective: these viruses, by definition, do not routinely and predictably contaminate the product. Only through unpredictable circumstances, virus might contaminate a given manufacturing run. In this case, the virus number and identity are not knowable *a priori*.

8.4.3   Differences Between Model or Surrogate Virus and Wild-Type Virus

Quantifying process capacity for virus clearance requires scale-down virus clearance studies. Inasmuch as the process-specific wild-type virus contaminant may be either unidentified, not able to be cultured, or otherwise unable to be used, common practice is to test with a panel of surrogate[1] viruses ("relevant" or "model" viruses) that represent ranges of physical and biochemical attributes. Suggestions on the identity of these viruses are found in ICH Q5A and are summarized in Table 8.2. Relevant and model viruses may differ in properties that affect virus clearance. Because there has not been a systematic comparison of the common relevant and model viruses, the differences are not always well understood. It is well known, however, that virus shape, aggregation state, and exterior surface charge influence virus removal, in particular for virus filtration [4]. Significant differences in susceptibility to virus clearance may exist even between genetically related viruses. For example, it has been shown that B19 virus, a relevant parvovirus for plasma products, is more susceptible to some inactivation methods than the commonly used model parvovirus MMV (mouse minute virus) [5].

8.4.4   Virus Spike Preparation Quality

Commonly, high-titer, laboratory virus preparations are inoculated into a process intermediate test article and the spiked fluid is subjected to the virus clearance step. Issues related to virus spike quality can confound one's ability to reproduce test results with the accuracy needed to derive reliable data for statistical analysis. Specific virus spike quality issues include variability in virus stock titers, existence and variability in the amount of host cell impurities, presence or absence of stabilizing agents such as BSA,

---

[1] ICH Q5A describes relevant viruses as those that are either identified as, or are the same species as, viruses that are known or likely to contaminate the cell substrate or any other reagents or materials used in the production process. A model virus is one that is closely related to the known or suspected virus (same genus or family). For biotechnology processes, model viruses with differing properties are used to characterize manufacturing process capacity.

TABLE 8.2. Examples of Virus Commonly Used for Clearance Studies

Virus Contaminant	Test Virus
RVLP (retrovirus-like particles)	X-MuLV
Adventitious RNA virus, nonenveloped	Reovirus 3, EMC
Adventitious RNA virus, enveloped	VSV, parainfluenza, Sindbis, BVDV, XMuLV
Adventitious DNA virus, nonenveloped	SV40, CPV, PPV, MVM
Adventitious DNA virus, enveloped	Pseudorabies

and application-specific interaction between the virus stock and the active ingredient or impurities in the test article.

8.4.5 Scale-Down Model Systems

Scale-down model systems are commonly used in viral clearance evaluation and validation studies. Commonly, it is the only time when a unit operation is actually challenged with a virus load to assess clearance capabilities. It is therefore imperative that the scale-down models accurately represent the performance of their cognate full-scale unit operations. Issues related to heat transfer, mass transfer, and fluid flow dynamics challenge one's ability to achieve fully parallel performance between small-scale and full-scale unit operations. Therefore, translation of data into knowledge should be done with a firm understanding of the limitations of the small-scale test system.

8.4.6 Inability to Confirm Manufacturing-Scale Effectiveness

Building from the small-scale processing knowledge, processes are scaled up incrementally and the overall CQA profile of the product is assessed. Process adjustments during scale-up are required often due to the imperfect relationship between small-scale and full-scale unit operations. With most chemical, physical, and biochemical quality attributes, the product may be assayed directly. With virus contaminants, assays of in-process intermediate and final bulk are of limited value due to limits of detection, as described subsequently. This places a heavier reliance on the scale-down model system.

8.4.7 Assay Limitations

There are three main categories of virus assays. The detection limit of each is relatively low because each assay analyzes only a small fraction of the full product volume. Each has major limitations with respect to one or more assay attributes:

Cell culture-based assays are performed by adding a small volume of test sample (containing virus) to a tissue culture and quantifying the interaction between virus and cells. This provides an indirect estimate of the number of virus particles in the original sample. The precision of this kind of assay is generally accepted as being $+/- \ 0.5 \log_{10}$ [3], high enough to cause difficulty in the interpretation of DOE-based study results. These assays also tend to be relatively slow and expensive.

Electron microscopy (EM) suffers from issues of specificity. First, species identification of a virus through EM is not reliable. Second, it is commonly not possible to determine if a virus particle has been inactivated, as the visual difference between active and inactive virus may not be always apparent.

Nucleic acid technology (NAT) assays include variations on polymerase chain reaction (PCR) or reverse transcriptase-polymerase chain reaction (RT-PCR). The benefits of these assays are their exceptional speed and low cost when compared to cell culture-based assays. PCR techniques employ virus-specific probes to ensure signal specificity, which allows study designs in which multiple viruses are spiked at one time [6]. Quantitative PCR (Q-PCR) allows one to quantify virus clearance with sensitivity approximately 100 times that of a conventional cell-based infectivity assay. Real-time Q-PCR allows one to quantify clearance during the time course of the experiment, not simply at the beginning and the end of the experiment. Still, PCR experimental design must account for issues related to endogenous viral DNA in preinactivation samples (background signal) and the possibility that a virus inactivation mechanism does not lead to a change in PCR signal.

8.5 DEVELOPING A VIRUS CLEARANCE DESIGN SPACE

The technical issues outlined above must be acknowledged. Nevertheless, the industry should not wait for their complete resolution. With the premise that a good design space exists within every robust process, design spaces can be approximated with today's technology and understanding of biology, chemistry, and engineering principles. One may segment the construction of a design space into six key elements (Fig. 8.1), some of which may be executed in parallel and some of which find a parallel to current practices and regulatory guidance.

8.5.1 Define Product Virus Safety-Related CQAs

A manufacturing process has the overarching aim of reliably meeting product CQAs. Thus, the process design space is predicated on the definition of the product design space. It has been acknowledged that "the design space approach to process definition is frequently hampered by inadequate information regarding product quality needs, that is, the deliverables (product specification)" [7]. With this dynamic as a backdrop, some biopharmaceutical manufacturers seek to document the maximum possible viral clearance level from the process. However, regulatory guidance does not stipulate an absolute log reduction value (LRV) requirement. To this point, Appendix 5 of the ICH Q5A guideline is sometimes interpreted as a mandate to achieve a 6 $\log_{10}$ safety margin per dose for endogenous retroviral particles, when in fact this appendix is intended to illustrate how to perform the calculation of estimated virus particles per dose.

To design in an appropriate level of viral clearance, one must establish target LRVs for the various models or relevant viruses. Risk assessment tools can aid in this determination [8]. The more common questions to consider are the following:

- What is the starting load and identity of virus?
- Are there specific patient risk issues (e.g., immunocompromised patient population)?
- What will the cumulative dose be (e.g., in chronic disease states)?
- What are the regulatory implications to not achieving a "high" level of viral clearance?

8.5.2 Define Viral Load

It is important to understand the potential viral load to apply QbD to virus clearance. Several scenarios illustrate this point. Consider a cell line with an unknown history of exposure to adventitious virus (e.g., a university source). With little understanding of the virus risk (virus type and number) that this cell line represents, one would be forced to set a very stringent process CQA (*i.e.*, a "high" LRV) for virus clearance. Extensive testing of that cell line may characterize the risk and counter the need to achieve an excessive level of viral clearance. Even more favorable is the use of a well-characterized cell line that has a good history of use in one's company. When using a well-characterized cell line in a new biopharmaceutical process, the viral load may be approximated through one's historical understanding of the cell line as that cell line has been used to express similar products. Even with this platform manufacturing strategy, however, it is necessary to confirm the virus load and that viral clearance targets are met.

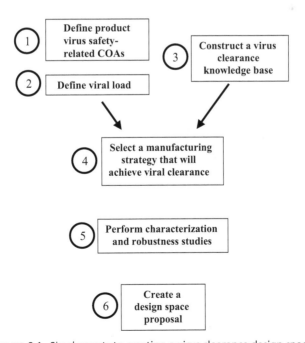

Figure 8.1. Six elements to creating a virus clearance design space.

Even with established and extensively tested cell lines, unexpected events can endanger the marketing of a product. For example, squirrel monkey retrovirus (SMRV) found in several cell lines was used to produce protein biotherapeutics [9]. This contamination was detected fortuitously during expression analysis of recombinant cell lines. A similar problem arose with the finding that MMV was introduced adventitiously into a manufacturing process [10]. This led to more stringent requirements for clearance of nonenveloped, resistant viruses.

Finally, during the 2002 West Nile virus (WNV) epidemic in the United States, it was learned that WNV could be transmitted via transfused blood. Plasma fractionators reviewed their production processes and validation data for clearance of a model virus, bovine viral diarrheal virus (BVDV), which is similar to WNV. It was confirmed that existing virus validation data could support continued use of the plasma products within the existing manufacturing space [11]. The manufacturing processes included inactivation steps such as pasteurization, solvent/detergent treatment, and vapor heating. Eventually, WNV was isolated and used in spiking studies to confirm an appropriate level of viral clearance.

8.5.3   Construct a Virus Clearance Knowledge Base

Over the last decade or so, considerable information on viral safety has been published in the form of peer-reviewed articles, conference proceedings, regulatory guidance, summary basis of approvals, and EPARs (European Public Assessment Reports). [12,13] Contract testing laboratories are another useful source of information for designing viral clearance into a process. These laboratories have valuable insights on robustness and current acceptance of the various viral clearance techniques. Multiple sources of information should be used to initiate and develop a viral clearance knowledge base for inactivation and removal mechanisms. The following provides a sampling of the kind of data and understanding that one may find in the public domain for various established viral clearance technologies.

Inactivation. In one survey, data were compiled for several model viruses. Parvovirus, reovirus, and hepatitis A were selected as models for nonenveloped viruses. Murine leukemia virus and pseudorabies virus were selected as models for enveloped viruses [14]. Table 8.3 describes properties of these model viruses.

For the parvovirus models, irradiation was found to be very effective under appropriate conditions, but heat-inactivation effectiveness varied with virus. Reovirus was inactivated by heat treatments of $\geq 60°C$, UV radiation, and gamma radiation. Hepatitis A was inactivated by irradiation and by dry heat, although dry heat effectiveness was affected by moisture content. Although several of the nonenveloped viruses presented a significant inactivation challenge, the enveloped viruses were shown to have a low resistance to physicochemical treatments. Heat and low pH have been demonstrated as effective inactivation mechanisms for enveloped viruses.

Although effective treatments were identified for each specific model virus, viruses exhibited significant differences in terms of their susceptibility to the various inactivation methods. In addition to the virus classes, many other variables affect the outcome and

TABLE 8.3. Properties of Some Model Viruses

Virus	Description[a]	Diameter (nm)	Shape	Physicochemical Resistance
Parvoviruses[b]	SS DNA, nonenveloped	18–24	Icosahedral	Very high
Reovirus	DS RNA, nonenveloped	60–80	Spherical	Medium
Hepatitis A	SS RNA, nonenveloped	22–30	Icosahedral	High
XMuLV, MuLV	SS RNA, enveloped	80–110	Spherical	Low
Pseudorabies	DS DNA, enveloped	120–200	Spherical	Medium

[a]SS, single-stranded; DS, double-stranded.
[b]Mouse minute virus (MMV), porcine parvovirus (PPV), canine parvovirus (CPV), and bovine parvovirus (BPV)

interpretation of viral inactivation studies. Among these variables are protein concentration, presence of product stabilizers, temperature (e.g., during pH inactivation), and exposure time. Finally, variability in the purity, viability, and titers of stock virus preparations can contribute to variation in virus clearance results. On a positive note, the industry is moving cooperatively toward a better understanding of the impact of virus spikes on clearance studies [15].

Removal. In biopharmaceutical manufacturing, virus removal is typically accomplished by virus filters and/or chromatography.

CHROMATOGRAPHY. Regulators have expressed concern that virus LRV provided by a chromatography column might deteriorate with repeated use of the column, necessitating resin lifetime studies for viral clearance. A study carried out at U.S. Food and Drug Administration specifically addressed consistency of retrovirus removal during repeated use of Protein A resins [16]. The authors used a PCR assay for reverse transcriptase (TM-PERT) as a surrogate for actual infectivity assays to evaluate retrovirus clearance. This study demonstrated that protein A column performance, measured by antibody step yield and breakthrough, deteriorated prior to noticeable reduction of retrovirus LRV. Similarly, a study on Q Sepharose™ Fast Flow operated in flow-through mode demonstrated that band spreading, reduced DNA clearance, and increased column backpressure occur prior to LRV loss for the nonenveloped SV40 and enveloped X-MuLV [17]. This study also demonstrated the importance of resin cleaning. Both the use of cleaning solutions that degrade the resin and not cleaning the resin decreased the ability of the column to remove SV40, MMV, and X-MuLV.

Generic and matrix approaches for viral clearance by chromatography were described many years ago [18]. In another publication [19], robust viral clearance ($>4 \log_{10}$) by anion exchange chromatography was reported for both SV40 and X-MuLV over defined conditions that include the following:

- pH range: 7.0–8.0
- Conductivity: 4.6–12 mS/cm
- Resin capacity: <200 g/L

- Flow rate: ≤ 500 cm/h
- Bed height: 11 cm.

The studies discussed here enable approximation of a design space for virus removal by chromatography.

VIRUS FILTRATION. Virus-removal filters operate primarily by a size exclusion mechanism. Virus filtration, especially for larger viruses, is considered to be a robust technique, that is, the size separation mechanism works regardless of virus type, feed solution, and operating conditions. However, commercially available virus removal filters vary from one filter manufacturer to another. They may differ in membrane materials, surface chemistries, membrane thickness, pore structure, and pore size distribution, which may lead to differences in virus retention, protein passage, and adsorption properties [20]. They may also be operated by normal flow filtration or cross-flow filtration.

During cross-flow filtration, a protein concentration polarization layer may form on the membrane surface, which may be favorable in terms of virus retention. On the contrary, during normal flow filtration, one may experience a flux decline over the course of the filtration. This may be associated with LRV decline that ranges from negligible to severe. This effect varies widely among filters and may be process specific.

During the development of a virus filtration step, one should consider, among other factors, the effects of protein concentration, aggregate concentration (exacerbated by aging and freezing of the feed), flux rate, and buffer chemistry. Recognizing that today's filters have inherent performance variability, one should evaluate the effect of membrane lot on filter performance. These studies should be done on scale-down filter models whose performance relative to full-scale filters has been characterized. In all of these studies, the two main outcomes required are successful product recovery and virus clearance.

8.5.4 Select a Manufacturing Strategy that will Achieve Viral Clearance

Generally, a manufacturing strategy is developed to achieve product CQAs related to physical and chemical attributes (e.g., purity, aggregates, and pH). Certain purification unit operations, such as chromatography and low pH elution from a protein A column, will be identified to have proven capability in achieving virus clearance. Additional unit operations designed solely for virus clearance (e.g., virus filtration) may be chosen to reach the final clearance target.

The complexity of selecting a manufacturing strategy will vary widely with the specific purification requirements of the drug product under consideration. More challenging situations involve lesser proven manufacturing practices or expression systems or less well-established categories of biopharmaceutical drug. These situations limit the extent to which a manufacturer may take advantage of the technical understanding and comfort level associated with existing technologies and practices. Accordingly, the burden of knowledge generation for these new approaches is likely to be relatively high.

On the other end of the spectrum are monoclonal antibodies, the most common example of a product category for which many manufacturers have adopted a manufacturing platform approach. A platform manufacturing process is a standardized process that a manufacturer establishes for all products of a like product category, for example, monoclonal antibodies. The established platform will have proven virus clearance capabilities. It is important, however, to ensure that the efficiency of the platform for viral clearance is confirmed for each specific product. The EU draft guideline on monoclonal antibodies states "The 'platform manufacturing' process will never be identical for each monoclonal antibody and interference by the product cannot be excluded beforehand. Therefore each process should be separately validated for its ability to remove viruses. However, for a new product, the manufacturer may partly rely on viral validation data obtained with other products manufactured with the same 'platform manufacturing' process. Such data may be considered supportive but the manufacturer will need to justify the relevance of the data and demonstrate that virus validation data obtained from the new product is comparable to data obtained for other products" [21].

Biopharmaceutical products vary in the extent to which they can tolerate virus clearance conditions. At the extreme, viral therapeutics are intrinsically difficult to handle, and virus safety assurance relies on complementary mechanisms that include extensive testing of raw materials and cell substrate and even end-product testing. Most biopharmaceuticals are susceptible to denaturation or aggregation, so one must assess product or process-intermediate stability, especially during exposure to harsh inactivation conditions. On the contrary, process impurities may actually protect a virus from inactivation. Preliminary scoping studies may reveal stability issues caused by either inactivation or removal steps. Purity and impurity profiles, potency, and product yield are some relevant measurable parameters.

When designing in a viral clearance strategy, a cost analysis is warranted. For typical biopharmaceutical processes, virus clearance steps are strategically placed to minimize costs. Consider, for example, the cost of a virus filter step or a low-pH inactivation step. Since the area of the virus filter is determined by the volume to be processed, one might want to place the filter into the process where the volume is reduced. Likewise, handing large volumes of low-pH inactivation and subsequent neutralization solutions can be costly. It has been observed that a reduction in buffer consumption by one-third can reduce downstream process costs by approximately 6% [22]. Although strategic, cost-conscious placement of viral clearance steps is appropriate, risks associated with product source and segregation of pre- and postviral clearance steps may dictate a less cost-effective placement in the process flow scheme.

8.5.5 Perform Characterization/Robustness Studies

After evaluating the viral clearance options and constructing a process strategy, viral clearance studies are needed to confirm that the clearance options are effective. Over the last decade, there have been more requests from regulatory agencies to test "worst-case" conditions for one or more independent variables. In some situations, "worst case" is clear. For example, as dwell time is an important factor for successful inactivation, inactivation kinetics are generally evaluated as part of a viral clearance study. However, the term "worst case" is often applied to, for example, high protein load, but it is not clear that a high load is always the worst case. The same is true for other independent variables (e.g., conductivity and pH) and the interactions between independent variables.

In a QbD approach, the manufacturer would run a series of virus clearance studies containing a strong element of statistical design of experiments (DOE), which would allow the quantification of main effects and the more likely two-way interactions. The outcome would be a statistical and perhaps mechanistic process understanding that, along with the product CQA requirements and the knowledge base of viral clearance technologies, forms the basis of a design space. In this context, Table 8.4 provides an overview of the typical significance of several common independent variables on both inactivation and removal.

The concept of "bracketed generic clearance" was described in 2003. In a controlled study, it was observed that an LRV of $\geq 4.6 \log_{10}$ of rodent type C retrovirus was achieved,

TABLE 8.4. Significance of Various Product, Process, and Testing Variables on Virus Inactivation and Removal

	Inactivation	Removal Chromatography	Virus Filtration
Product			
pH	++	++	+
Conductivity/ionic strength	++	++	+
Temperature	++	++	+
Buffer composition	++	++	+
Feedstream purity	++	++	+
Aggregate concentration	+	+	+
Protein concentration	++	++	++
Type of product	+	+	+
Process			
Unit operation time	++	+	+
Flux or flow rate	NA	++	++
Pressure	NA		+
Cleaning	NA	++	NA
Reuse	NA	++	NA
Temperature gradient/heat transfer	+	NA	NA
Mixing efficiency	+	NA	NA
Virus testing			
Virus spike purity	++	++	++
Virus spike viability	++	+	+
Virus spike volume	++	++	++
Virus titer	++	++	++
Availability of scale-down model	++	++	++
Virus class	++	++	+
Virus size	+	+	++

++, very significant in most cases; +, significant in some cases; and NA, not applicable.

TABLE 8.5. An Outline Viral Clearance Design Space for Inactivation of Type C Retrovirus

Attribute	Setting
pH	≤ 3.8
Incubation time	≥ 30 min
Incubation temperature	$\geq 14°C$
Buffer system	Citrate or acetate
Total protein concentration	≤ 40 mg/mL
NaCl concentration	≤ 500 mM
pI of protein	3–9
Process intermediate	Cell-free intermediate after initial capture chromatography step
Product	Monoclonal antibody (not retrovirus targeted)

provided the conditions in Table 8.5 were met [23]. Although not discussed in QbD terminology by the authors, this study has in effect outlined a viral clearance (in this case inactivation) design space for a specific virus particle. Because processes are evaluated on a case-by-case basis, regulatory agencies to date have not yet accepted this generic bracketing concept for either clinical materials or for license applications.

Mammalian virus clearance studies run under GLP are currently required to validate viral clearance into a specific process. The cost of these studies is high, so it is not feasible to construct a full design space with this kind of study. To manage this issue, several strategies may be employed. First, one may consider use of nonhazardous, nonmammalian virus models (*i.e.*, bacteriophages) as a means of generating a large amount of clearance data with relative speed and cost-effectiveness. Second, one may generate preliminary mammalian virus clearance data under non-GLP conditions. One biopharmaceutical company has described the use of non-GLP facilities to perform a comparison of new and classical viral clearance methods [24]. Others rely on contract testing organizations that can perform preliminary studies without the rigors of GMP compliance to provide valuable process design information. Third, one must account for the possibility that future process changes will lead to the need for new studies. This is especially the case for early phase studies, for which most companies opt for using center-point-only study design.

Finally, planning is essential. A well-characterized scale-down clearance model must be developed before the virus testing may begin. Experimental issues related to, for example, virus viability and virus assays during virus clearance studies should be anticipated and discussed with the test facility. Product shipping studies should be executed to assure that the product arrives at the test facility free of excessive aggregation or denaturation: aggregates may lead to over- or underestimation of viral clearance, depending on the clearance mechanism. Planning can mitigate risks of project delays or of finding out too late that the manufacturing process provides insufficient viral clearance.

8.5.6 Create a Design Space Proposal for the Unit Operation

From the experimental data derived above and learning from the scientific literature, one derives knowledge about the impact of critical process variables on viral clearance

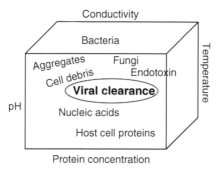

Figure 8.2. Conceptualized multidimensional process design space within which all contaminants and impurities are effectively cleared.

effectiveness. Presumably, similar knowledge will have been generated to define a space within which all of the biological, chemical, and physical CQAs will be met (Fig. 8.2). The final design space would combine the knowledge gained from all of these studies, thus ensuring that all CQAs are reproducibly achieved. The design space would then be proposed, through regulatory submission, as representing the boundaries within which one would manufacture a biopharmaceutical product.

When describing a design space, one should account not only for manufacturing process parameters, but for ancillary operations as well. As an example, consider cleaning of a chromatography column. Changes to the cleaning frequency or cleaning reagent concentration in manufacturing may alter the virus clearance profile (virus capacity, separation from the protein peak, etc.) of the unit operation. One should also assess fermentation process windows. It has been observed, for example, that when the metabolic state of the cells or rates of protein expression change, retrovirus synthesis may increase up to 2 $\log_{10}$ [25].

A design space is described as a multidimensional window of operation that is sufficiently well characterized to allow processing anywhere within that window. Anticipating that comprehensive virus clearance DOE designs may be prohibitively resource intensive, it is possible that initial viral clearance design spaces may comprise relatively few dimensions. Nevertheless, these initial efforts will represent important advances in the industry's collective efforts to adopt QbD.

8.6 STAYING IN THE DESIGN SPACE

A basic outcome of the QbD and design space approach is greater flexibility to document and execute postapproval changes. With added flexibility will come the added care to ensure that proposed change is justifiably represented by the existing design space. Changes that fall outside the design space must be avoided.

One should consider with caution the addition of viral clearance steps after a process has been designed. While augmenting virus safety, the added step may compromise other product or process CQAs. Conversely, process changes apparently unrelated to virus clearance may have a direct and unintended impact on virus clearance.

Staying within the viral clearance design space will require communication among functions such as process development, validation, and manufacturing. Collectively, these functions will possess the information required to make the correct decision to maintain regulatory compliance and product safety.

8.7 CONCLUSION

QbD and design space implementation challenges the industry to develop and present a strong knowledge of the virus clearance capabilities of their manufacturing processes. Very few biopharmaceutical manufacturers have the resources to build comprehensive virus clearance design space understanding on their own. Therefore, the change to a QbD and design space philosophy also challenges the regulatory community and the industry to enter partnerships to augment collective understanding in the field of virus clearance. That collective understanding may be enhanced by the following measures:

- Compilation of existing data, such that the industry can transform data into a knowledge base for the application of virus clearance technology. This effort may also uncover issues that require further understanding.
- More academic studies that further elucidate mechanisms of viral clearance for technologies for which that information is still limited.
- Studies to establish levels of equivalence among viruses. Particularly beneficial could be the identification of bacteriophages that show similarity to mammalian viruses with respect to susceptibility to virus clearance technologies.
- Cooperative efforts aimed at creating standardization in areas such as virus assays and virus spike quality attributes.

Currently, the industry has the required technical tools and regulatory guidance to incorporate well-controlled virus clearance unit operations into biopharmaceutical manufacturing processes. As more information is shared in the biopharmaceutical industry and knowledge bases are augmented, the collective ability within the industry to design and control more well-characterized processes will increase. The benefits of increasingly stronger process characterization and understanding are mutually shared among the industry, regulators, and patient populations.

ACKNOWLEDGMENTS

The polymerase chain reaction is covered by patents owned by Roche Molecular Systems and F. Hoffman-LaRoche. A license to use the PCR process for certain research and development activities accompanies the purchase of certain reagents from licensed suppliers. Sepharose is a trademark of GE Healthcare companies, GE Healthcare Bio-Sciences AB, a General Electric Company, Bjorkgatan, Uppsala, Sweden.

REFERENCES

[1] Juran JM. *Juran on Quality by Design. The New Steps for Planning Quality into Goods and Services.* New York: Simon and Schuster; 1992.

[2] ICH Q6A. Specifications: Test Procedures and Acceptance Criteria for New Drug Substances and New Drug Products: Chemical Substances. 2000. Available at www.ich.org.

[3] ICH Q5A. Viral Safety Evaluation of Biotechnology Product Derived from Cell Lines of Human or Animal Origin. 1997. Available at www.ich.org.

[4] Korneyeva M, Rosenthal S, Charles P, Trejo S, Pifat D, Petteway S. Important parameters for optimal target virus clearance: virus spike issues. *Am Pharm Rev* 2004;7(1):38–42.

[5] Kasermann F, Kempf C, Boschetti N. Strengths and limitations of the model virus concept. *PDA J Pharm Sci Technol* 2004;58(5):244–249.

[6] Valera CR, Chen JW, Xu Y. Application of multivirus spike approach for viral clearance evaluation. *Biotechnol Bioeng* 2002;80(3):257–267.

[7] PDA Letter. October, Bethesda, MD, 2007.

[8] ICH Q9. Quality Risk Management. Available at www.ich.org. 2005.

[9] Middleton PG, Miller S, Ross JA, Steel CM, Guy K. Insertion of SMRV_H viral DNA at the c-myc gene locus of a BL cell line and presence in established cell lines. *Int J Cancer* 2002;52:451–454.

[10] Garnick RL. Raw materials as a source of contamination in large-scale cell culture. *Dev Biol Stand* 1998;93: 21–29.

[11] Kreil TR, Berting A, Kistner O, Kindermann J. West Nile virus and the safety of plasma derivatives: verification of high safety margins, and the validity of predictions based on model virus data. *Transfusion* 2003;43:1023–1028.

[12] U.S. Food and Drug Administration. http://www.fda.gov.

[13] EMEA. http://www.emea.europa.eu/htms/human/epar/eparintro.htm.

[14] Sofer G, Lister DC, Boose JA. Part 6, inactivation methods grouped by virus. *BioPharm Int* 2003; (Suppl.):S-37–S-42.

[15] PDA Task Force on Virus Spike Standardization. Technical Report. In progress.

[16] Brorson K, Brown J, Hamilton E, Stein KE. Identification of protein A media performance attributes that can be monitored as surrogates for retrovirus clearance during extended re-use. *J Chromatogr A* 2003;989:155–163.

[17] Norling L. Impact of multiple re-use of anion exchange chromatography media on virus removal. *J Chromatogr A* 2005;1069:79–89.

[18] U.S. Food and Drug Administration. 1997. Points to Consider in the Manufacture and Testing of Monoclonal Antibody Products for Human Use. Bethesda, MD. Available at www.fda.gov.

[19] Curtis S. Generic/matrix evaluation of X-MuLV and SV40 clearance by anion exchange chromatography in flow-through mode. Oral Presentation at IBC's 8th International Conference on Process Validation for Biologicals, March 7–8, San Diego, CA; 2005.

[20] Parenteral Drug Association. PDA Technical Report No. 41: Revised 2008 Virus Filtration. *PDA J Pharm Sci Technol* 2008;62(5–4). Available at www.pda.org.

[21] CHMP Draft Guideline on Production and Quality Control of Monoclonal Antibodies and Related Substances. EMEA/CHMP/BWP/157653/2007. Available at www.emea.europa.eu.

[22] Sofer G, Jagschies G. Economy by design for biopharmaceutical chromatographic processes. *Pharm Tech Europe* 2007;19(12):36–40.

[23] Brorson K, Krejci S, Lee K, Hamilton E, Stein K, Xu Y. Bracketed generic inactivation of rodent retroviruses by low pH treatment for monoclonal antibodies and recombinant proteins. *Biotechnol BioEng* 2003;82(3):321–329.

[24] Zhou JX, Dehghani H. Viral clearance: innovative versus classical methods: theoretical and practical concerns. *Am Pharm Rev* 2006;9(5):74–81

[25] Brorson K, deWit C, Hamilton E, Mustafa M, Swann PG, Kiss R, Taticek R, Polstri G, Stein KE, Xu Y. Impact of cell culture process changes on endogenous retroviral expression. *Biotechnol Bioeng* 2002;80(3):257–267.

9

APPLICATION OF QUALITY BY DESIGN AND RISK ASSESSMENT PRINCIPLES FOR THE DEVELOPMENT OF FORMULATION DESIGN SPACE

Kingman Ng and Natarajan Rajagopalan

9.1 INTRODUCTION

The goal of pharmaceutical development is to design and establish a formulation composition and its manufacturing process to consistently meet all the appropriate quality attributes required for its intended therapeutic performance and safety profile. In the past, unacceptable product quality due to either component or process variability was only detected from laboratory testing of the finished product. Recently, FDA has challenged the pharmaceutical industry to embrace the concept of Quality by Design (QbD), where the appropriate level of quality must be designed into the product and the process based on scientific understanding and quality risk management approach. In particular, the information and knowledge gained from pharmaceutical development studies and manufacturing experience provide scientific understanding to support the establishment of the design space, specifications, and manufacturing controls [1].

Development of an acceptable dosage form for biotechnology-derived products is extremely challenging because protein molecules require a unique three-dimensional structure for biological activity. In addition, proteins are composed of numerous reactive groups that can be subject to various types of degradation chemistry such as covalent and noncovalent aggregation, deamidation, cleavages, oxidation, and denaturation, all of

Quality by Design for Biopharmaceuticals, Edited by A. S. Rathore and R. Mhatre
Copyright © 2009 John Wiley & Sons, Inc.

which may lead to loss of biological activity. Due to the complex physicochemical characteristics of proteins, the only technically viable dosage forms are in most cases parenterals. These dosage forms are usually aqueous solution and suspension formulations or lyophilized powders that are reconstituted in a well-defined diluent system immediately before injection. The manufacturing process for parenteral solutions employs typical unit operations whereas the suspension may require more complex unit operations. The formulations are filled in a container/closure system (e.g., vials, prefilled syringes, and cartridges) and dosed using a delivery device (e.g., syringes and injectors).

To have quality built into the product by design, the development scientists need to understand and fully characterize the inherent stability properties of these complex and reactive protein molecules during the design and development of the dosage form, formulation composition, container closure system, and manufacturing process. Among these aspects, the formulation composition needs to be established first, where the relationship between the critical stability properties and the formulation variables has been identified according to risk assessment [2], optimized and evaluated for shelf life requirements, defined and summarized in the context of formulation design space [1]. The design of the formulation composition should also, to some extent, take into consideration the selection, justification, and control strategies of other drug product components such as the properties of the active pharmaceutical ingredient (API), excipients, container closure system, and manufacturing process. This chapter aims at providing an overview of QbD principles and risk assessment approaches with emphasis on the development of formulation design space using the example of a monoclonal antibody. The development of manufacturing process design space and application of process analytical technology (PAT) will not be discussed here.

9.2 QUALITY BY DESIGN (QbD) APPROACH

The QbD approach for formulation development starts with a clear definition of the target product profile (TPP) in accordance with patient needs related to dosing, convenience, and compliance, as well as marketing requirements for the intended therapeutic indication. Specific formulation requirements and attributes pertinent to the TPP, for example, the need to have a single use sterile stable solution formulation, need to be determined prior to the start of commercial development. For biotechnology-derived products, a significant hurdle to successfully develop a stable solution formulation is mostly derived from the multiple degradation pathways and modifications located in the different regions of the molecule. Extensive preformulation and characterization studies with proper analytical and bioassay support are conducted to first understand the types and location of the various chemical modifications, physical instability challenges, and the impact on the biological activity relevant to the target therapeutic indication. In particular, the effect of the various formulation-related factors such as pH, temperature, headspace, trace metals, excipients, and so on needs to be fully characterized. A preliminary stability risk assessment, based on the molecular degradation knowledge, is used to establish those degradations that are likely to have the greatest risk of impacting

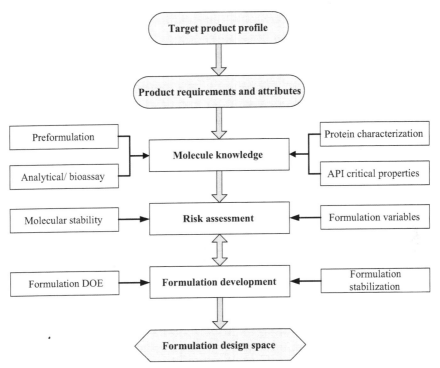

Figure 9.1. Schematic of the formulation design space development via QbD and risk assessment principles.

the critical quality attributes (CQAs). In addition, an initial risk assessment based on preformulation knowledge is also needed to establish the criticality of those formulation factors that are likely to influence those modifications identified to have high stability risk. The risk assessment is essential for designing the appropriate multivariate experiments to define the formulation design space by systematically optimizing the critical formulation parameters with respect to stability properties that have a high risk of impact on the CQAs. As a result, a commercial composition that meets the TPP requirements can be defined based on the product stability response knowledge derived from the formulation design space. A schematic summarizing the development of formulation design space via the QbD and risk assessment approach is shown in Fig. 9.1. The subsequent sections in this chapter provide further description and discussion for each step according to the schematic as shown in Fig. 9.1 when applied to the case of monoclonal antibody.

9.3 TARGET PRODUCT PROFILE (TPP)

To build and develop "quality" into the final product, the first step is to understand the commercial product definition with respect to therapeutic indication, patient dosing,

TABLE 9.1. Summary of Product Requirements and Attributes Pertinent to Develop Formulation Design Space

Product Requirements	Product Attributes
Parenteral dosage form via subcutaneous dosing	Sterile ready to use solution
Single use for patient self-administration	Simple disposable delivery device such as prefilled syringe
24 months shelf life stability under refrigerated storage conditions	Stable solution formulation
Injection considerations such as volume and needle size	Prefilled syringe with suitable needle size
	Viscosity of solution formulation is acceptable and not a limiting factor for syringeability

compliance, and marketing requirements. Some examples of the product requirements and attributes, and hence pertinent to formulation development, are summarized in Table 9.1.

Based on initial evaluation of these product requirements and corresponding attributes as per the target product profile, the type of dosage form, for example, aqueous solution formulation, and suitable container closure system and manufacturing process can be proposed.

9.4 MOLECULAR DEGRADATION CHARACTERIZATION

An important step to develop a stable solution formulation is to identify the "degradation hot spots" in the molecule through a combination of stress methods to induce and force the structural and chemical modifications. The aim of the forced degradation studies is to understand how sensitive different degradation pathways are to external stress and other factors. The stress methods include temperature, pH, buffer effects, and oxidation susceptibility with respect to peroxide, trace metals, oxygen, headspace, and light. These studies provide important information to help identify the likely degradation pathway and intrinsic stability properties of the molecule. Different types of degradation products can be monitored by various analytical methods to assess the ability of the protein drug molecules to withstand these stresses. The data from these studies are critical to understanding the stability properties, to determine which analytical methods indicate stability, and are used to aid in the design of preformulation studies.

In the case of a monoclonal antibody that is composed of two heavy chains and two light chains joined to form a "Y-shaped" molecule, it is important to understand the molecular structure and activity relationship. A schematic of antibody structure is shown in Fig. 9.2 [3].

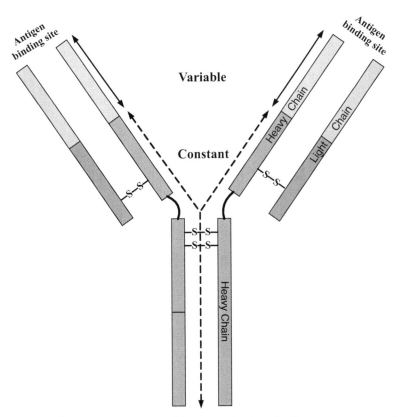

Figure 9.2. Schematic of the monoclonal antibody structure.

The variable region, which includes the ends of both heavy and light chains, is the antigen binding site. The variable region is further subdivided into hypervariable (HV) and framework (FR) regions. The HV regions contain a high ratio of different amino acids in a given location and are directly responsible for the antigen binding activity. The FR regions have more stable amino acid sequences and form a beta sheet scaffold structure to hold the HV regions in position to contact the antigen. On the other hand, the constant regions do not directly participate in the antigen binding activity. Identifying the type of chemical modifications and the associated molecular regions with respect to the antigen binding activity helps design the preformulation studies and determine stabilization strategies. An example summary of these types of results for an antibody example is provided in Table 9.2. In this case, the forced degradation results for a particular antibody under development may indicate that the variable region is susceptible to aggregation at acidic pH, oxidation at elevated temperature, and in the presence of trace metals, whereas the constant region is more susceptible to deamidation at alkaline pH region and oxidation at elevated temperature. These results are critical to the

TABLE 9.2. An Example Summary of Forced Degradation Results

Mode of Forced Degradation	Summary of Results
Temperature	Variable region
	Significant effect on oxidation
	Constant region
	Significant effect on deamidation and oxidation
	Moderate effect on soluble aggregate formation
pH	Acidic pH
	Significant effect on soluble aggregate formation
	Alkaline pH
	Significant effect on deamidation in constant region
Trace metals, peroxide, and light	Variable region
	Significant effect on cleavage and oxidation
	Constant region
	Significant effect on oxidation
	Moderate effect on soluble aggregate formation
Headspace	No significant effect on stability properties

understanding of stability properties and are used to aid in the optimization and selection of the solution formulation.

9.5 ACTIVE PHARMACEUTICAL INGREDIENT (API) CRITICAL PROPERTIES

A critical prerequisite for developing a stable solution formulation is to understand the critical properties of the API. Upstream cell culture and/or other processing steps may generate a variety of chemical modifications. The stability growth potential of these modifications could cause significant solution instability that cannot be easily corrected during formulation development. In particular, there could be large lot-to-lot variability of these chemically modified forms during API process development. If this is discovered early in development, significant purification efforts should be devoted to focus on the removal and/or proper control of these chemically modified forms during API process development. Therefore, it is critical to assess solution stability of multiple API lots with an aim to establish a correlation between API process and subsequent formulation development studies and drug product stability properties. Short-term accelerated stability study should be conducted on multiple lots of API to assess and benchmark critical solution stability properties during formulation development and optimization. In particular, it is extremely useful to establish a correlation between the timing of API lot history and the various formulation development studies, such as forced degradation, preformulation, formulation design of experiment (DOE), and prototype stability studies. As a result, the risk of potentially significant lot-to-lot variability of API critical properties is minimized and provides some assurance that formulation development results are representative. Drug product prepared from future API lots would be expected to perform similarly.

9.6 PREFORMULATION CHARACTERIZATION

The aim of preformulation studies is to characterize the effect of the various formulation variables on different types of degradation and modification. The effect of pH on solution stability is probably the most important factor. In particular, effects of pH should be evaluated for (i) physical and chemical stability, (ii) thermal stability, and (iii) solubility. Preformulation studies should be conducted early to understand the relative stability of the molecule over a wide pH range. Different protein stability challenges are expected at different pH regions. For example, aggregation and denaturation are usually more prevalent at acidic pH whereas deamidation and disulfide scrambling are more pronounced near alkaline pH. Solubility is also an important factor, where protein solubility is usually at a minimum near its isoelectric point. Lower solubility typically leads to lower physical stability resulting in aggregation and precipitation challenges. As a result, it is critical to balance between chemical stability, physical stability, thermal stability, and solubility to identify the optimal pH range for proper dosage form development. If an optimal pH region does not exist for a feasible solution formulation, a freeze-dried formulation needs to be considered. Alternatively, specific excipients can be added to address specific instability issue and therefore enhance the probability of identifying an optimal pH region (see below for an example of physical instability). Once the optimal pH region has been identified, the effect of buffer also needs to be characterized. Depending on the pH condition, several common buffer types, such as acetate, citrate, phosphate, Tris, and histidine, should be studied.

Because proteins are folded into a higher order structure for biological activity, they are commonly susceptible to aggregation as a result of physical stress of shear and shaking. Physical instability is often a concern and may be more difficult to control than chemical instability. In particular, a high degree of hydrophobicity can lead to more susceptibility to physical instability in the presence of additives that impart ionic strength, such as sodium chloride. A common approach to resolve the physical instability is through the use of surfactants. Polysorbates 20 and 80 are commonly used surfactants that are highly effective stabilizers. Different types of agitation studies that can simulate the potential stress on the protein molecule both during manufacturing and shipping/storage should be included in the preformulation studies to identify the need of a surfactant stabilizer and the appropriate concentration to protect against physical instability. In some cases, the addition of a surfactant stabilizer can also expand the pH range of acceptable physical stability. Figure 9.3 shows an example of balancing multiple analytical properties to identify the optimal region of pH. Each arrow in Fig. 9.3 represents the acceptable pH range for the particular analytical property. In this particular case, an optimal pH region does not exist that balances chemical and physical stability properties. However, addition of a surfactant as stabilizer can effectively expand the acceptable pH region for physical stability, which is shown as the gray arrow in Fig. 9.3.

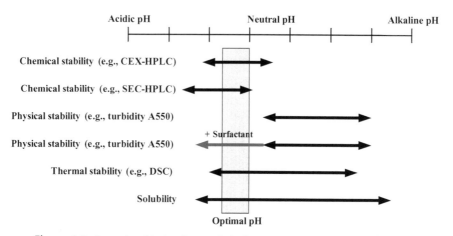

Figure 9.3. Example of balancing analytical properties to identify optimal pH.

9.7 INITIAL FORMULATION RISK ASSESSMENTS

As stated earlier, a significant hurdle to successfully develop a stable solution formulation is mostly derived from the multiple degradation pathways and chemical modifications located in different regions of the molecule. A preliminary stability risk assessment should take into consideration the impact of these modifications on the critical quality attributes directly linked to the TPP, as well as the stability growth potential.

In the case of a monoclonal antibody, modifications located in the variable region (Fig. 9.2) known to have impact on the activity are considered to have high stability risk and need to be characterized and carefully monitored during stability studies. For those chemically modified forms without stability growth potential, the risk factor should be initially assigned as medium until the stability trend is fully characterized later. Chemical modifications in the constant region with no impact on the binding activity and no stability growth potential are considered to have low risk. However, a medium risk factor is assigned when there is stability growth potential where the rate constants need to be monitored to finalize the risk factors. Another important consideration is aggregation, which is considered to be of high risk due to the potential impact on the drug product performance with respect to the binding activity and immunogenicity effect. In addition, aggregation usually grows on stability and hence needs to be carefully characterized and monitored during stability studies. The type and location of each modification, as well as the potential impact on critical quality attributes and associated stability risk, are summarized in Table 9.3.

A second element of the initial formulation risk assessment investigates possible impact of formulation parameters and conditions, such as pH, excipients, trace metals, light, and so on, on solution formulation stability. Based on preformulation and characterization experience, as well as common protein chemistry and parenteral knowledge, the factors assessed are shown in the cause–effect Ishikawa (Fishbone)

TABLE 9.3. Example of Stability Risk Assessment

Modification/Degradation	Activity Impact	Grow on Stability	Risk Assessment
Variable region	Yes	Yes	High
	Yes	No	Medium
Constant region	No	Yes	Medium
	No	No	Low
Aggregation	Yes	Yes	High

diagram (Fig. 9.4) [4]. The criticality of each of these identified factors can be determined based on forced degradation and preformulation study results to help set up the multivariate formulation optimization study.

9.8 FORMULATION OPTIMIZATION AND DESIGN SPACE

The next step is to optimize the critical formulation variables with respect to those degradations of high risk that are identified via risk assessment based on preformulation and forced degradation studies. A multivariate statistical DOE study is set up to optimize the key formulation parameters with respect to chemical and physical stabilities.

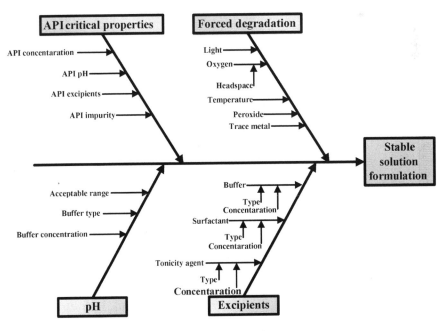

Figure 9.4. Example of the Ishikawa (Fishbone) diagram for stable solution formulation.

The goal of the DOE study is to develop a solution formulation suitable for commercial development. In particular, a quantitative model can be developed based on statistical analysis of the results. The DOE study can be set up to have multiple arms with different objectives. The core design covers the critical variables for solution formulation. Additional experiments can also be set up to address the effect of certain formulation variables beyond the DOE core design range and the effect of other specific excipients such as antioxidants or chelating agent. The response attributes are measured by a variety of analytical techniques to evaluate both chemical and agitation-based physical stability properties.

Statistical analysis of the DOE results provides rate constant and other analytical response prediction essential for assessing solution formulation stability properties as a function of the critical formulation variables. The detailed analysis and predicted contour profile of rate constants at 5°C for each individual stability property are first established. Figure 9.5 shows an example of the contour profile of predicted rate constants at 5°C for a chemical stability property. These contour plots clearly establish the response of the chemical stability rate constants in the entire range of critical variables 1 and 2 from the DOE study. Similar statistical modeling of the DOE results can also be carried out to model all the critical stability response as a function of critical variables 1 and 2.

After the statistical model has been determined for the response of each analytical property, a "formulation design space" can be established by combining the individual response to satisfy a collective set of stability criteria for the critical degradation properties identified to have high risk of impacting the CQA. The DOE results summarized in the context of formulation design space provide a comprehensive understanding of the response of the critical product attributes as a function of the

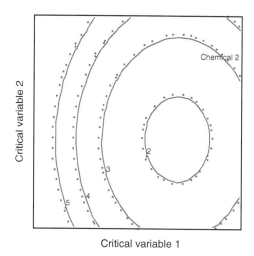

Critical variable 1

Figure 9.5. Contour profile example of predicted 5°C rate constant for a chemical stability.

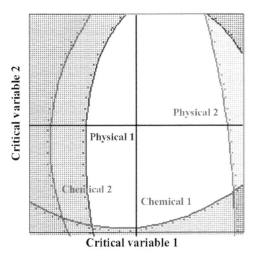

Figure 9.6. Concept of formulation design space. (See the insert for color representation of this figure.)

critical formulation variables. Furthermore, this information forms the knowledge base for selecting a commercial solution formation with sufficient robustness that is likely to meet the shelf life stability requirements. To illustrate the formulation design space concept, projection of the acceptable range of the two critical formulation variables that meet all the specified stability criteria is shown in Fig. 9.6. The size and shape of the "design space," which is depicted as the "white region" in Fig. 9.6, depend on the combination of different stability criteria projected for the various critical analytical properties.

9.9 SELECTION OF SOLUTION FORMULATION COMPOSITION

The solution formulation composition is defined by optimizing stability properties with respect to degradation pathways that have high risk of impacting the CQA. The selection is based on the DOE results and statistical modeling analysis, as well as preformulation characterization and forced degradation study results. In particular, experimental results are used to justify and mitigate the formulation risk factors identified with the Ishikawa diagram as shown in Fig. 9.4. For example, the rationale of selecting the solution formulation composition for a typical case example of antibody is summarized in Table 9.4.

The next step is to place the proposed commercial formulation on stability at multiple temperature conditions to evaluate the effect of long-term real-time storage. To support potential dosing levels for later stage clinical studies, stability studies are set up over a range of concentrations. In addition, different combinations of container closure systems are also included to delineate any potential incompatibilities. The real-time

TABLE 9.4. Rationale of Solution Formulation Selection

Factors	Selection Rationale
API critical properties API Concentration ⟶ API pH ⟶ API excipients ⟶ API impurity ⟶	The same optimal and robust pH range, with identical choice of buffer type and concentration, is selected to match the final drug product Critical excipient such as surfactant stabilizer may need to be included in the API matrix to minimize physical instability issues
Forced degradation Light ⟶ Oxygen ⟶ Headspace ⟶ Temperature ⟶ Peroxide ⟶ Trace metal ⟶	The recommendation for appropriate storage temperature condition is based on preformulation characterization, forced degradation, formulation DOE optimization, and stability study results to meet the product shelf life requirement The recommendation of proper measure for headspace and light effect is based on forced degradation and preformulation characterization results Other excipients, such as antioxidant and chelating agent, may need to be included depending on formulation stability properties
pH Acceptable range ⟶ Buffer type ⟶ Buffer concentration ⟶	The acceptable pH range is selected to balance the chemical stability, physical stability, thermal stability, and solubility properties. The optimal and robust pH range can be determined by DOE results summarized in the context of formulation design space Appropriate buffer type and concentration is selected to have adequate buffer capacity and minimal effect on solubility and stability properties within the optimal and robust pH range

TABLE 9.4. (*Continued*)

Factors	Selection Rationale
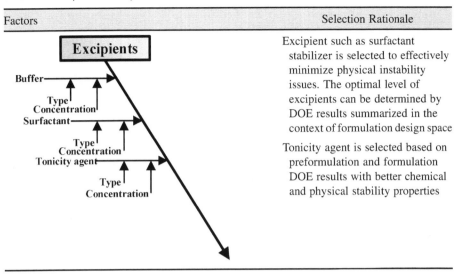	Excipient such as surfactant stabilizer is selected to effectively minimize physical instability issues. The optimal level of excipients can be determined by DOE results summarized in the context of formulation design space
	Tonicity agent is selected based on preformulation and formulation DOE results with better chemical and physical stability properties

prototype stability results are compared to the statistical model prediction to update and finalize the stability risk assessment.

9.10 SUMMARY

The design of a commercial formulation composition in accordance with the Quality by Design principles generally involves the following steps:

- Identification of target commercial drug product profile.
- Preformulation and forced degradation studies to characterize molecular stability properties, impact of formulation variables, and other factors.
- Preliminary stability risk assessment with emphasis on direct impact on the activity based on preformulation and forced degradation studies results.
- Initial formulation risk assessment to establish the cause–effect relationship of different factors and solution formulation stability via the Ishikawa (Fishbone) diagram.
- Multivariate DOE studies to optimize the formulation composition and define a robust design space to meet the expected shelf life of 24 months at 5°C.
- Establish formulation design space based on DOE results and stability properties projections.
- Select commercial solution formulation based on design space, molecule knowledge, and risk assessment.

ACKNOWLEDGMENTS

The authors would like to thank Vincent Corvari and Mike DeFelippis for a useful discussion on QbD approach and critical reading of the chapter, Bryan Harmon for useful discussion on stability risk assessment, and Suntara Cahya for useful discussion on statistical concepts of experimental design.

REFERENCES

[1] ICH Q8. Harmonised Tripartite Guideline Pharmaceutical Development, 2005.

[2] ICH Q9. Harmonised Tripartite Guideline Quality Risk Management, 2005.

[3] van Dijk MA, Vidarsson G. Monoclonal antibody-based pharmaceuticals. In: Crommelin DJA, Sindelar RD, editors. *Pharmaceutical Biotechnology. 2nd ed.* London: Taylor & Francis; 2002. p 283–299.

[4] Tague NR. *The Quality Toolbox. 2nd ed.* Milwaukee: ASQ Quality Press; 2004. p 247.

10

APPLICATION OF QbD PRINCIPLES TO BIOLOGICS PRODUCT: FORMULATION AND PROCESS DEVELOPMENT

Satish K. Singh, Carol F. Kirchhoff, and Amit Banerjee

10.1 INTRODUCTION: QbD IN BIOLOGICS PRODUCT DEVELOPMENT

The inherent complexity of biologics presents a challenge in the application of QbD and design space concepts. The link between product/molecule attribute and clinical performance is not always well understood. The heterogeneity of the product makes it difficult to precisely measure all the attributes of the molecule. It can be difficult to define the quality attributes (QAs) that are truly critical to safety and efficacy. Consequently, the focus shifts toward using clinical experience and maintaining process consistency. Similarly, the functional relationship between process parameters and quality attributes may not be readily defined.

The development of a stable formulation for a biologic therefore involves accounting for the inherently complex structure of the protein and the multiple degradation pathways it may undergo. A compromise among competing degradation pathways may be required to develop an optimized formulation (e.g., acid hydrolysis and deamidation). A large body of knowledge has been generated that can aid the formulation scientist in this task, and a number of summaries and reviews have been published (see, for example, Refs [1–3]). For biologics, heterogeneity of the bulk drug substance is the rule rather than exception, and the upstream and downstream processes are a major source

Quality by Design for Biopharmaceuticals, Edited by A. S. Rathore and R. Mhatre
Copyright © 2009 John Wiley & Sons, Inc.

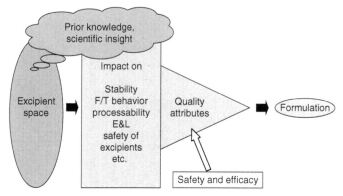

Figure 10.1. Application of the QbD concept to formulation development.

of this heterogeneity. A well-designed formulation is, therefore, key to controlling final product variability and ensuring an active product over the entire shelf life.

A rational process of developing a formulation is summarized in Fig. 10.1 and fits the tenets of QbD as well as ICH Q8. The excipient space represents the choice of excipients the formulator has available. This excipient space, combined with prior knowledge as well scientific insight gained from preformulation (hot spot analysis, pH-stability profile, etc.) and biophysical characterization studies (such as CD, DSC, and fluorescence), allows the formulator to design appropriate formulation development studies. These studies screen the test formulations through "filters" such as storage stability, freeze/thaw behavior, processability, safety of excipients at dosed levels, extractables, leachables, and so on. The objective is to select formulation(s) that are "optimal" with respect to the quality attributes of the biologic.

Similarly, an illustration of the QbD process to an aseptic fill/finish has been exemplified in Fig. 10.2. A successful and robust fill/finish operation will incorporate process parameters as well as material quality attributes into the analysis. An assessment of the impact of these on the process performance and product quality can be performed. This will lead to the identification of those process parameters and material quality attributes that may require further studies to develop the relationship to process performance and product quality attributes.

Usually, when discussing QbD or design space, the discussion automatically shifts to process variability. For a biologic, the drug product manufacturing operation itself tends to be simple and is generally considered not to have an impact on product quality. However, a systematic and comprehensive risk assessment of the product and the process allows for the identification of areas where further experimentation will lead to improved process and product knowledge. The quality of the product is a consequence of the process as well as the formulation—a holistic approach is required. This chapter provides an example of the risk assessment process and presents some case studies on how this analysis can lead to the development of design space for both formulation and process.

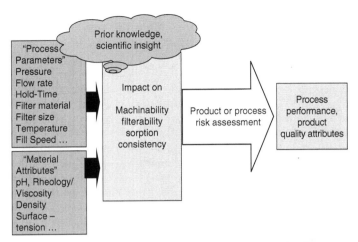

Figure 10.2. Application of the QbD concept to process development.

10.2 RISK ASSESSMENT PROCESS

To align with ICH Q8, Q9, and Q10, a comprehensive risk assessment process was developed in Pfizer: "Right First Time for Co-Development" (RFT). The objective of the work process is to apply risk-based approaches to generate process understanding and process control that will create (1) regulatory flexibility and (2) the foundation for continuous improvement (life cycle). The work process is intended to be continually improved for efficiency and effectiveness. The implementation of such a process should lead to decrease variability in manufacturing, faster change implementation, scientific knowledge at hand to support quality investigations, reduce cost of goods, and create innovation.

10.2.1 Process Understanding

Inputs to the process control variability of the output. The inputs (x) are process parameters such as people, equipment, measurements, process, materials environment, and so on while the outputs (y) are protein content, aggregates, host-cell protein, biopotency, endotoxins, sterility, peptide map, contaminants, and product degradants. The objective is to establish the functional relationship between quality attribute (y) and process parameter(s) (x), that is, $y = f(x_1, x_2, x_3 \ldots)$.

10.2.2 Objectives

The objective of the process is to gain agreement on process scope, decide what is important to evaluate, prioritize parameters based on risk, gain agreement on high-level experimental strategy, and identify and prioritize process analytical technology (PAT) applications.

10.2.3   Work Process

The RFT work process consists of following four broad steps:

(1) Risk assessment
(2) Experimental planning
(3) Prioritization
(4) Experimentation

This is a repetitive process usually carried out at various stages of the development, generally starting prior to the manufacture of ICH lots, prior to process validation lots, and continues through postlaunch.

The Risk assessment steps consist of first creating a process map and identifying "focus areas," followed by identification of quality attributes and how they are measured for each of the focus areas (FAs), then identifying process parameters using a cause and effect matrix. The scores from the cause and effect matrix are generally used to prioritize the parameters and then define the experimental approach.

The remainder of this chapter will be used to illustrate these steps.

10.3   EXAMPLES

10.3.1   Identification of Focus Areas, Process Parameters, and Quality Attributes

This is the first task in the risk assessment involving creation of a process map and identification of process parameters and quality attributes.

Identification of Focus Areas (FA).   The drug product manufacturing process can be viewed as a progression of specific unit operations and defined as a process map. The specific unit operations can be labeled as focus areas. Typically, drug product manufacturing begins with compounding, but may require manipulation of the drug substance (active pharmaceutical ingredient), especially if a biologic is used. Compounding is directly followed by filtration, filling, and finishing. Additional unit operations such as component preparation and in-process testing are inherent to drug product manufacturing. Together, these make up the process map.

Figure 10.3 outlines an example of a process map containing focus areas that may be identified for a liquid drug product. Many biologics are formulated at the last step of the downstream purification process. In certain cases, minor dilution steps may remain, but in general, the process of drug product manufacture primarily involves aseptic fill/finish operations, (followed by lyophilization in approximately 50% of the products).

In the case considered here, a frozen drug substance requiring further dilution is used. FA1 is identified as the thaw of the drug substance. Component preparation, including manufacturing equipment, vials, stoppers, and overseals is identified as FA2. FA3 is compounding or combining the drug substance with the other excipients required in the final drug product. In-process testing is usually conducted during the compounding

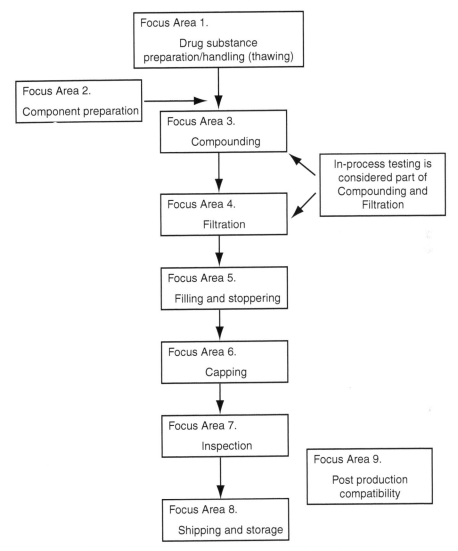

Figure 10.3. An example of a drug product process map.

(FA3) and filtration (FA4) unit operations. FA5 is filling and stoppering. FA6 is capping followed by inspection (FA7) and shipping/storage (FA8). Postproduction compatibility, such as in-use stability, is included in this process as FA9.

After the identification of the focus areas, the entire process map is complete. Each focus area is then detailed by the specific activities occurring in that unit operation. Consider compounding, FA3. The compounding unit operation consists of weighing each component, adding them in order, mixing, and confirming pH and concentration of the active. Confirmation of pH and concentration is part of in-process testing. Figure 10.4 delineates the compounding step or FA3 in our case study.

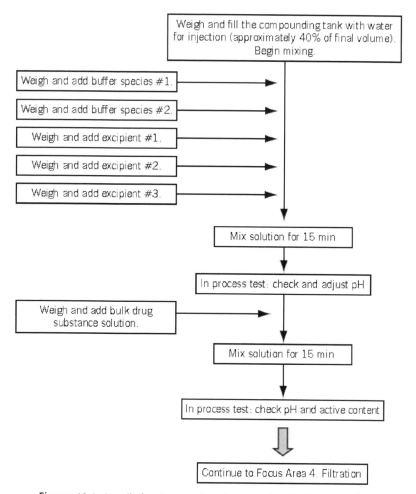

Figure 10.4. Detailed unit operations in Focus Area 3: Compounding.

Identification of Process Parameters. A process parameter is defined as a process input. These are typically machines, materials, measurements, processes, people, and environments. In our case study for FA3, compounding process parameters identified include: mixing speed, mixing time, hold time, temperature, excipient quality, excipient handling, balance accuracy, and operator training. These are outlined in the risk assessment sheet in Table 10.1.

For fill and finish activities (FA5), the process parameters would include filling speed, needle bore size, needle height, vial size, fill volume, temperature, and line speed along with people and environment-related parameters.

Identification of Quality Attributes. A quality attribute is defined as a process output. Quality attributes are typically physical, chemical, or microbiological properties

TABLE 10.1. QbD Risk Assessment Scorecard Example for Focus Area 3

| | Quality Attribute/Rank | | | | | | | | | | | | |
| | 10 | 7 | 7 | 10 | 10 | 7 | 7 | 7 | 10 | 10 | 5 | | |
Process Parameter	Concentration of Active	Deamidation	Oxidation	Aggregation	pH	Color	Clarity	Fragmentation	Content Uniformity	Bioactivity	Process Time	SCORE	Experimental Strategy
Temperature	1	10	10	10	1	7	7	10	1	10	1	543	DOE1
Excipient weight accuracy	1	7	7	10	10	1	7	5	1	7	7	514	DOE1
Drug substance solution pH	1	10	1	7	10	5	1	10	1	5	7	464	DOE1
Hold time of bulk drug substance	1	7	7	7	1	5	5	10	1	7	10	458	DS stability
Drug substance weight accuracy	10	1	1	1	7	5	1	5	1	10	10	431	FMEA1
Mixing Time	1	5	5	5	1	1	5	1	10	5	10	389	OFAT1
Mixing Speed	1	1	5	5	1	1	5	1	10	5	7	346	OFAT1
Excipient handling	1	1	10	5	1	1	1	5	1	5	1	261	FMEA2
pH meter accuracy	1	1	1	1	10	1	1	1	1	1	10	225	FMEA3
Operator/analyst training	1	1	1	5	1	1	1	1	7	1	5	210	on-the-job-training
Excipient quality	1	1	5	1	5	5	1	1	1	1	1	186	FMEA4
Order of excipient addition	1	1	1	1	1	1	1	1	5	1	1	130	FMEA5

or characteristics of a material or process. Care needs to be taken to include attributes that can be experimentally measured.

Commonly identified quality attributes of a biologic product include color, clarity, subvisible particulates, pH, concentration, aggregation, oxidation, deamidation, bioactivity, and sterility. In this case study of a liquid biologic drug product, the primary quality attributes of the compounding step are identified as concentration of active, deamidation, oxidation, aggregation, pH, color/clarity, fragmentation, content uniformity, bioactivity, and process time.

10.3.2   Ranking Quality Attributes (QA) and Parameters

After completing the process map, delineating the unit operations in each focus area and identifying the process parameters and quality attributes, the next step is to rank the quality attributes and parameters for each focus area.

The process parameter rank scores are assigned as follows, in relation to a specific quality attribute:

The process parameters for the compounding steps (FA3) were ranked as mixing speed $= 7$, mixing time $= 7$, hold time $= 10$, temperature $= 10$, excipient quality $= 5$, excipient handling $= 7$, balance accuracy $= 10$, and operator training $= 5$ (see case study Risk Assessment sheet in Table 10.1).

QA scoring takes into account prior knowledge, experience, and expectations. For example, the current emphasis on aggregation and its potential impact on immunogenicity suggest that aggregate levels in a product will be a high-ranked QA. Similarly, it is possible that the level of oxidation in a biologic has no impact on its potency/activity, efficacy or safety, the case in which oxidation QA may be scored lower. Oxidation in this case is an attribute that could demonstrate control from a process and product consistency point of view in this case, but would have a lower QA score from a compounding perspective.

Quality attributes are ranked similarly with the following:

- Score of 10 is given to those attributes where it is established or expected that a direct relationship to product quality or safety (including manufacturing safety) exists
- Score of 7 is given to attributes where it is unsure, but an impact to product quality or safety or key business drivers is expected;
- Score of 5 is given to attributes that are unlikely to impact product quality or safety
- Score of 1 is given to attributes where no product quality or safety impact is expected.

The quality attributes for the compounding steps (FA3) were ranked as concentration of active $= 10$, deamidation $= 7$, oxidation $= 7$, aggregation $= 10$, pH $= 10$, color/clarity 7, fragmentation $= 7$, content uniformity $= 10$, bioactivity $= 10$, and process time $= 5$.

10.3.3   Prioritizing Experiments and Experimental Strategies

Scoring is the next step in the process. By combining the process and the quality attribute ranks from the risk assessment table (Table 10.1) along with the scoring for each parameter/quality attribute, a total score for each parameter is calculated. Typically, there is a natural break in the totals. The highest scores are considered highest priority, medium scores are considered medium priority, and so on. The experimental strategies to address the parameter/attribute are then defined.

Experimental strategies include failure mode and effects analysis (FMEA), one factor at a time (OFAT), science of scale (SoS), and design of experiments (DOE). Multiple experimental types may be used in a single focus area. The experimental strategy is selected to gather enough information to provide results that can sufficiently support the aim of the study. In the case of compounding (FA3), the highest ranking scored experiments were viewed as a statistically designed experiment (DOE). The factors of temperature, excipient and active concentration (amount from weighing accuracy), and pH of the product were tested simultaneously to capture interactions between these parameters and the effect on product quality.

10.3.4 Experimentation

Case Study DOE: Formulation Robustness and Formulation Design Space. The risk assessment scorecard in Table 10.1 identified excipient weight accuracy and pH as two process parameters with a high-risk score, and a DOE strategy was selected to address this. Formulation development studies define an optimal composition that represents a point in the formulation space, but exact compositions can be difficult to produce in large-scale processes. For example, preparation of large volumes of buffers by weighing water and buffer salts may not yield the exact composition as specified by the formulator. It is therefore important to understand the robustness of the formulation to process-induced variations in composition parameters. The range of composition parameters that the biologic can accept without significant impact on all (critical and noncritical) quality parameters delineates the design space of the formulation. An example of such a study is given below. It may be noted that while a DOE approach to liquid formulation development has been commonly used (e.g., Refs [4–6]), robustness is not often taken into consideration.

For the case study discussed here, the composition for the biologic solution formulation identified from formulation finding studies is given in Table 10.2.

TABLE 10.2. Formulation Identified as Optimal for a Biologic

Component	Quantity (Molar Ratio Compared to Active)	Function
Active	1	Active ingredient
Buffer strength	20 mM	Buffer
PH	5.5	—
Excipient A	1657	Tonicity adjuster/stabilizer
Excipient 1 (Surfactant)	1.1	Stabilizer
Excipient 2 (Chelator)	2.0	Stabilizer
Water for injection	q.s.	Solvent

An assessment was made on the ranges to be tested on the basis of knowledge of process capability and knowledge gained from earlier studies as to the function and impact of the individual excipients on the biologic. Instead of individual salts, the buffer was considered as a single composition variable based on buffer strength, since once it is formulated, it is controlled by conductivity and pH limits. With buffer conductivity acceptance limits of 0.8–1.2 mS/cm, it was determined that a 15 mM or a 25 mM buffer would fall outside the limits and therefore represents an appropriate range for buffer strength. The range for pH was set between 5.2 and 5.8 based on earlier pH-stability studies. The ranges for excipients 1 and 2 were set at 0.6–1.7 and 0.6–3.4 molar ratio of excipient to protein, respectively, based upon previous experience and preformulation studies. A range for excipient A was not evaluated since the specification range for osmolality would identify any anomalies in the level of the tonicity adjuster, and the amount was considered robust for the function as stabilizer due to its large molar excess compared to the active.

A simple half factorial experimental design with four factors and center point was conducted as shown below in Fig. 10.5. Formulations were prepared and placed on

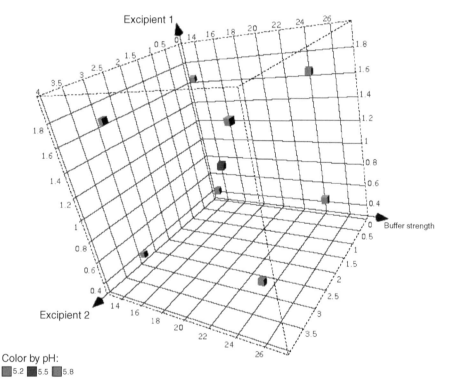

Color by pH:
5.2 5.5 5.8

Figure 10.5. Experimental design for robustness of product composition. The center point represents the optimal or target composition. (See the insert for color representation of this figure.)

stability under preferred (2–8°C) and accelerated (25°C) storage conditions in the appropriate container/closure system. The data available after a year were analyzed as an intermediate point. It was found that at 2–8°C, none of the quality attributes were impacted, suggesting a robust formulation. A design space for the formulation could therefore be defined. Under accelerated conditions, the analysis showed impact of pH and buffer strength on deamidation and oxidation (see Fig. 10.6a and b). The model dependencies on buffer strength are weak. Impact of pH on deamidation is well understood, and the finding fits with knowledge of the deamidation mechanism. The interaction plots confirm that pH 5.5 is optimum with respect to deamidation. Low pH also favors lower oxidation.

Case Study FMEA: Process Analysis. FMEA is a tool to analyze a process or product to determine potential reliability problems through identification and detailing of a process map that leads to the finished product, what might go wrong at the various steps in this process, how might it go wrong, and what would be the impact. An extended analysis requires an assessment of the (1) S = severity of the mode of failure and relates to the effect on the product safety or efficacy (2) O = probability that the failure mode will occur, and (3) D = difficulty to detect the mode of failure. The three aspects of the failure mode are assigned numbers, usually between 1 and 10, with the higher the risk, the greater the number. These numbers are multiplied to obtain the risk priority number (RPN = O × S × D). Understanding the risk level through the RPN can then help in prioritizing activities, experiments, and plans for risk mitigation strategies. FMEA can be applied to all aspects of product development and manufacture (Table 10.3).

An example of such an analysis pertaining to formulation excipients is provided in Table 10.4. The process of analysis is exemplified here by considering the first failure mode. Incorrect amounts of charged excipients were identified as a potential mode for failure under the category excipients. The potential effect of incorrect excipient levels would be an out of specification (OOS) at release or a stability failure. Of the potential causes for such a failure, simple incorrect addition was assigned a low RPN. The probability of occurrence that this would not be detected was considered low due to the presence of GMP manufacturing controls. The severity was assigned a 5 since the failure mode could cause a batch failure but was not likely to harm patients. The action required in this case is to ensure that batch records are correct and the proper controls are in place. The other potential cause for this mode of failure was considered to be simply variability/errors in weighing large amounts of material and liquids as a consequence of accuracy of balances, and so on in the relevant ranges. The probability that this would occur was assigned a 6 and the probability that this would not be detected was assessed as 8. An RPN of 240 implied that some corrective/remedial action would be required—in this case by identifying the range of excipients level that would be acceptable. From the first row, it is clear that the variability of the amount of excipients charged amounts to a high RPN. One of the ways to address this failure mode would be the robustness and formulation design space experiment shown as Case Study DOE in the previous subsection.

The high RPN number for the second failure mode identified here pertains to the quality of raw material. Experience and insight suggest that among the formulation ingredients, excipient 1 is most susceptible to deterioration in quality, which may have an

(a)

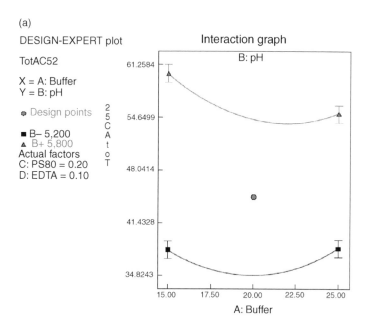

(b)

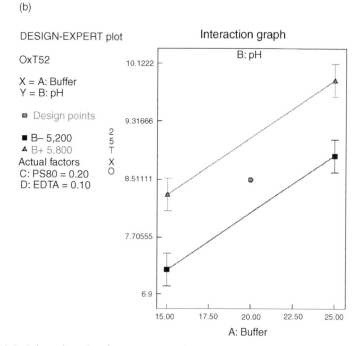

Figure 10.6. Selected results of composition robustness case study from Fig. 10.5. (a) Deamidation as a function of pH and buffer strength. (b) Oxidation as a function of pH and buffer strength.

TABLE 10.3. Some Aspects of Product Development and Manufacture Where FMEA can be Applied

Formulation Design	Filling (Machine, Process)
Excipients	Container/closure and machinability
Formulated bulk	Stopper and capping
Storage	Visual inspection
Filter selection	Secondary packaging
Sterile filtration	Cleaning validation

impact on the product (e.g., through aggregation or oxidation). To address this failure mode, two studies were designed. The first examined the impact of storage conditions on the quality parameter of excipient 1; most expected to have an impact on product quality. In this case, it was taken to be the oxidation status of the excipient, as measured by the peroxide number. It was determined that the facility where the excipient is used obtains it in 3 kg bottles. Each run uses 0.5 kg of excipient 1 and a campaign usually consisted of 4–6 runs. A small DOE study was conducted to determine the impact of storage temperature (refrigerated versus room temperature) and air head-space (low, high) on the rate of oxidation of the excipient. While the generally accepted practice is to store the excipient under refrigerated conditions, a drawback is the possibility of moisture absorption due to repeated opening and closing of the bottle when cold. Similarly, while layering with nitrogen is the accepted practice, the quality of the nitrogen layering in a manufacturing environment in actual practice may be questionable when there are no controls placed on the process and the quality of the layering is up to the thoroughness of the operator. The study provided a correlation between change in oxidation status of the surfactant with time as a function of the storage temperature and headspace volume. This was followed by another simple DOE in which the product was formulated with excipient of varying levels of oxidation, the range of which was determined in the previous study. These formulation compositions were then placed on long-term and accelerated stability and a correlation created between the product quality parameters and the surfactant quality and thereby to the storage condition of the raw material. The outcome was a limit on the acceptable oxidation number of the excipient and a recommendation to the handling of an opened excipient bottle in the plant (a hold-time operational range)— "Use within 30 days of opening. Discard bottle if not used within 30 days."

An example of an FMEA applied to a portion of the fill/finish process is provided in Table 10.5. To address this failure mode, sorption studies for both excipient 1 and protein were performed on the filter to ensure the filter is appropriate (Fig. 10.7). Another study was done to assess impact of filtration/fill stop on the material when the solution is stationary in the filter (Table 10.6). These laboratory studies confirmed that the filters were appropriate and the degree of sorption was measurable. Other experiments showed that degree of sorption was dependent on flow rate and filter size, leading to a decision to study these phenomena at scale. Experiments at scale confirmed that the sorption over the range of process flow rates was acceptable and the surfactant concentrations obtained fell in the range that had been defined acceptable in the formulation composition robustness study. The machine stop study allowed a process parameter range to be set for this

TABLE 10.4. An Example of an FMEA Application to the Drug Product Characteristic "Excipient"

Characteristic (Main Branch)	Potential Failures	Potential Effects	Potential Causes	O	S	D	RPN	Action Required
Excipients	Not in right amount	OOS/stab failure	Variability during manufacture in weighing large volumes	6	5	8	240	Assess acceptable ranges. Identify critical excipients to control
			Incorrect amount added	1	5	1	5	Batch records
			Not in spec	1	5	1	5	QC control
	Not of right quality	OOS/stab failure	Aged raw material (RM)	8	8	5	320	Assess impact of range of spec on critical attribute of critical RM

TABLE 10.5. An Example of an FMEA Application to the Drug Product Process Characteristic "On-Line Filtration by N_2 Pressure"

Characteristic (Main Branch)	Potential Failures	Potential Effects	Potential Causes	O	S	D	RPN	Action Required
On-line filtration by N_2 pressure	Adsorption of excipient 1 during machine stop	Product not homogeneous excipient 1 concentration too low	Inappropriate membrane, long stand-time of solution in filter	5	8	8	320	Adsorption tests as part of filter selection; process robustness testing; simulated machine stop studies; process validation
	Adsorption of protein during machine stop	OOS assay result	Inappropriate membrane, long stand-time of solution	3	8	5	120	Adsorption tests as part of filter selection; process robustness testing; simulated machine stop studies; process validation

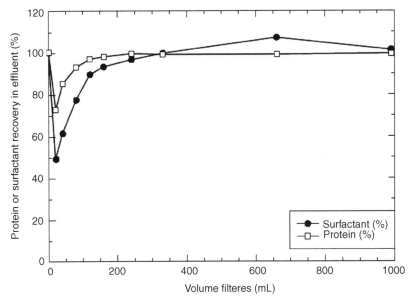

Figure 10.7. Results of filter sorption study to verify filter selection.

eventuality. In other situations, such a study has been used to define discard volumes after machine stoppage.

Case Study 3: Compatibility with Infusion Components. Section 2.6 of ICH Q8 states "The compatibility of the drug product with reconstitution diluents (e. g., precipitation, stability) should be addressed to provide appropriate and supportive information for the labeling. This information should cover the recommended in-use shelf life, at the recommended storage temperature and at the likely extremes of concentration. Similarly admixture or dilution of products to administration (e.g., product added to large volume infusion containers) might need to be addressed." This aspect of the drug product is placed in Focus Area 9 of the process map (Fig. 10.3).

TABLE 10.6. Assessment of Filling Machine Stoppage on Product Composition

Sampling Time	Cumulative Flow of Bulk (L)	Filter Pre-Pressure (bar)	Recovery Surfactant (% of Bulk)	Recovery Protein (% of Bulk)
Start of test	2	0.5–0.7	—	—
Before machine stop (after 0.3 h of process)	4	0.5–0.7	100	100
After machine stop (after 2.3 h of process)	4	—	95	100

TABLE 10.7. Infusion Component Materials of Construction Qualified for Use with Biologic Intravenous Injection

Normal Saline Bags	IV Infusion Lines	Catheter/Infusion Sets	Syringes/ Needles	Infusion Pumps
PVC/DEHP	PVC/DEHP	Catheter: polyurethane, silicone, stainless steel	Polypropylene, latex-free rubber, silicone, stainless steel	Peristaltic
PVC/DEHP	PVC/DEHP/ polyethylene (PE)	Catheter: teflon, silicone, stainless steel	—	Volumetric (cassette)
Polyolefin (polyethylene/ polypropylene)	PVC/TOTM (tri octyl trimellitate)	Infusion set: teflon, silicone, stainless steel, PVC (tubing)	—	Volumetric (shuttle)
—	Polyurethane (DEHP-free)	Polyurethane, silicone, stainless steel	—	—

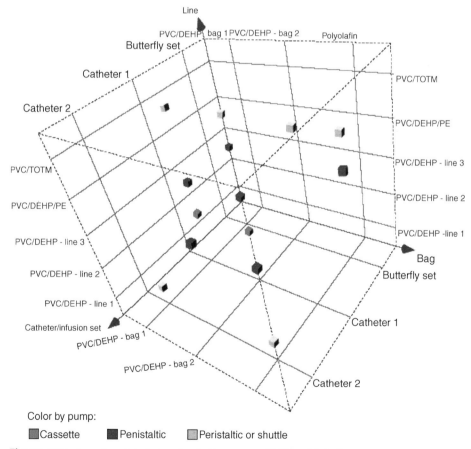

Figure 10.8. Experimental design for testing compatibility of infusion components with a biologic infusion product. (See the insert for color representation of this figure.)

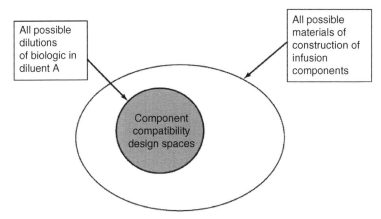

Figure 10.9. A schematic illustration of the component compatibility design space for an infusion product.

An important component of the above compatibility requirement is the qualification of various infusion components for use with the biologic product. For a product to be marketed worldwide, it is obvious that a large number of component manufacturers would be in question. Since it is not possible to qualify all potential components for a variety of manufacturers, a bracketing strategy based on DOE was applied. The strategy based on qualifying materials of construction instead of specific components makes the results applicable to components other than those tested. The elements tested in the DOE are listed in Table 10.7, along with the actual design in Fig. 10.8.

The study showed the compatibility of the biologic with the materials of construction of the most commonly used infusion components and the concentration range within which the compatibility existed, thus defining a design space in which the biologic could be used. This is illustrated schematically in Fig. 10.9.

10.4 CONCLUSIONS

A systematic work process has been defined to apply risk-based approaches to generate process understanding and align ICH Q8, Q9, and Q10. The examples provided illustrate the many ways in which the concepts of QbD and design space can be applied to biologic product—formulation and process development.

REFERENCES

[1] Carpenter JF, Pikal MJ, Chang BS, Randolph TW. Rational design of stable lyophilized protein formulations: some practical advice. *Pharm Res* 1997;14:969–975.

[2] Patro SY, Freund E, Chang BS. Protein formulation and fill-finish operations. *Biotechnol Ann Rev* 2002;8:55–84.

[3] Wang W, Singh SK, Zeng DL, King K, Nema S. Antibody structure, instability and formulation. *J Pharm Sci* 2007;96:1–26.

[4] Fransson J, Hagman A. Oxidation of human insulin-like growth factor I in formulation studies. II. effect of oxygen, visible light and phosphate on methionine oxidation and evaluation of possible mechanisms. *Pharm Res* 1996;13:1476–1481.

[5] Gupta S, Kaisheva E. Development of a multidose formulation for a humanized monoclonal antibody using experimental design techniques. *AAPS PharmSciTech* 2003;5:74–82.

[6] Srivastava A, Goldstein J, Agarkhed M, Pirrotta D, Rivera J, Wojcik R, Pan G, Tracy S, Zhou Q, Tarnowski J. Development of a freeze dried formulation to protect against non-enzymatic hinge region cleavage for a monoclonal antibody. Poster Presentation, National Biotechnology Conference, Boston, MA, AAPS, 2006.

11

QbD FOR RAW MATERIALS

Maureen Lanan

11.1 INTRODUCTION

Design of experiments (DOE) is one strategy used to simplify development of a biological manufacturing process. The experiments are performed systematically using combinations of factors and the results provide quantitative models that predict outcomes given the inputs. When multiple factors are included in the design, multivariate mathematical models are created that describe the "design space" for a process. Quality by design (QbD) is the process used to define these multidimensional equations and, ultimately, to translate these equations into an understanding of all the combinations of discreet process conditions that can be used to give similar results from a manufacturing process. Because it takes quite a few experiments to create and test these models, the process of comparing factors and selecting manufacturing conditions takes place during development, often in small scale, before manufacturing begins.

Once a process is transferred into a manufacturing plant, priority is shifted from mapping the design space to making consistent product with a well-controlled process. Fewer batches can be run in parallel because of cost and logistics. The emphasis on making product for use in patients takes precedent. It is during this transition to manufacturing that omissions or incorrect ranges chosen for the DOEs from development can be discovered; yet, it is also the time when the experimental tools like DOE are abandoned. While application of QbD can minimize the number and severity of surprises on transfer of a process to manufacturing, less attention is given to detect variables once a process is in manufacturing, opening up a possibility for trouble later on.

Raw materials may need a different strategy—one that continues from development into the manufacturing stage. The reason is that most time-dependent, lot-to-lot changes in raw materials happen during the manufacturing stage. Unlike with particular manufacturing steps, QbD applied to raw materials must provide strategies—from supply chain to cell harvest—to detect and manage changes in raw materials that may happen for the first time after the process is being run in manufacturing. So, raw material QbD could be viewed as an attempt to leverage learning derived from intentional variability applied during the development of a process into the uncharted design space mapped out by future raw material variation. Precisely because raw material changes can take place after development when systems are not designed to detect or learn from change, there may be a distinct role for QbD applied to raw materials.

Unlike physical process conditions, raw materials are not under the direct control of the manufacturer. They are usually made by external vendors. In addition, raw materials can vary in purity from lot to lot in ways that take a long time to become apparent compared to the time needed to develop a manufacturing process. Because of this long timescale associated with raw material variability, a QbD strategy for raw materials will look different from QbD for a process step. This chapter examines some of the characteristics of raw materials that make them distinct and provides an example to explore how QbD concepts might apply to raw materials. The data in this chapter were not generated as part of a QbD study, instead they are part of an investigation for a manufacturing process. The data may help illustrate how raw materials and feed streams could be incorporated into an overall QbD strategy. With the retrospective data, specific benefits and challenges that could result from "QbD for raw materials" will be explored.

11.2 BACKGROUND

The typical manufacturing process for a biopharmaceutical is a batch process with upstream cell-culture production, downstream purification, and finally drug product manufacture. Each stage has different critical aspects related to raw materials where a change in a raw material could change product or process consistency. While QbD could be applied to any of these stages, this discussion will focus on the upstream, cell culture. Media used for cell culture has so many chemical components that, in some ways, it can mimic the complexity of a cell itself. It is an extreme case for raw material control, needing both simplicity and more understanding about detailed composition. The nutrient formulations used in media for cell culture, even when simplified, can contain more than 40 compounds. In addition, reactions can occur between compounds in the media generating even more chemical complexity. When this potpourri of chemicals is used to grow cells, which add their own degree of complexity, the result can be a system that is operationally sensitive and prone to variability from run to run. Without a well-designed strategy to manage the numerous variables associated with cell-culture media, cell-culture development could be rate limiting and/or result in poor process capability once in the manufacturing plant.

11.3 CURRENT PRACTICE FOR RAW MATERIALS

Current practice for raw material analysis is described in ICH guide Q7: Good Manufacturing Practice Guide for active pharmaceutical ingredients (API). Materials used to prepare active pharmaceutical ingredients (both small molecules and biologics) need to have the identity of each batch confirmed on receipt and a Certificate of Analysis (C of A) provided from the supplier. In addition, the supplier must be qualified as suitable based on audits of their facility, their analytical results must be confirmed to be reliable, and a sampling plan is needed for each incoming material.

Even though these guidelines are followed, there remains a possibility that changes in raw materials will occur during the life cycle of a product and that some of these changes will not be measured in the analytical tests reported on the C of A. Examples of unanticipated raw material changes that have an impact on production come from both big pharmaceutical companies and biologic manufacturers. These include subtle changes in the level of impurities ranging from hydrogen peroxide affecting protein stability to particle size variations affecting dissolution rates. The incoming specifications and tests used to accept a raw material for use are based on the results from development studies. Once a process is in manufacturing, quality systems are put in place to ensure consistent practices, creating a momentum to maintain the status quo. Additional effort is needed to change a process once it is in manufacturing even in response to new information.

11.4 QbD IN DEVELOPMENT

DOE is applied extensively during the small-scale development stage of a process and then, to a lesser extent, through clinical drug manufacturing and into early commercial manufacturing. In a similar way, QbD for raw materials would ideally start during the development phase of a process by identifying which variables in a raw material will affect product yield or quality. Development experiments identify these key factors so that they can be controlled. At this stage, QbD for raw materials might be indistinguishable from QbD applied to develop a cell-culture process.

In contrast, while DOE is a key tool used for QbD in the development stage of a process, DOE is not normally incorporated into a manufacturing process itself. DOE is not part of the standard design practice for pharmaceutical or biopharmaceutical processes [1]. Since DOE is not planned into manufacturing processes, it cannot be used as a tool to detect or learn from future variation in raw material composition. While simplifying the manufacturing operation in the short term, this strategy does reduce the opportunity to continue to develop a resilient manufacturing process, one designed for ongoing learning. It is assumed that major sources of variation in raw materials are discovered and managed during small-scale development. The working assumption is that there is no need to have a separate raw material QbD strategy in manufacturing because DOE-based media development experiments are adequate.

During development, it is easier to introduce changes to a process. Information generated during development is likely to result in financial benefits simply because the result is more likely to be used. QbD for raw materials has been envisioned to date primarily as an aspect of early process development. For example, materials from different vendors are typically compared during small-scale development. This provides the opportunity to qualify a different vendor, a useful strategy to reduce risks should one supplier run into problems. Vendor and lot comparison during development is not new and could be considered as a long-standing example of QbD applied to raw materials.

A somewhat newer thrust of QbD for raw materials can be classified as deformulating media to build understanding about the impact of each media component on the cell culture. Deformulation can be done from an analytical or DOE approach and ideally occurs during small-scale process development. In practice, for some older mammalian cell-culture processes, natural products such as plant hydrolysates or serum are used in the media. The complexity of these mixtures is such that analytical platforms are only now becoming robust enough to be able to identify significant components from these complex mixtures over the long time frames characteristic of raw material fluctuations [2]. The example in this chapter shares experiences from a deformulation study. More typically, media composition is optimized during development based on extensive DOE in micro or small-scale bioreactors. The effort focuses on defining a minimal media composition and feed strategy to yield good growth. Information from these DOE experiments can also be used to create chemically defined media.

Another approach to minimize the impact of raw material variation on a process is to use chemically defined media. With chemically defined media, complex mixtures derived from natural sources such as cotton or wheat hydrolysates are replaced with combinations of chemicals. However, some obstacles still need to be overcome to make chemically defined media perform consistently. Perhaps the most basic is a reproducible small-scale bioreactor system to support designed experiments. In addition, the cost of goods for a chemically defined media could be higher than that for unrefined media ingredients. Part of the cost comes in the additional steps needed to purify the components at the vendor and additional steps to prepare solutions with more ingredients. Lastly, even when chemically defined media is available, it still may not be possible to anticipate the effect of future variations in raw materials based on development experience.

11.5 QbD IN MANUFACTURING

DOE suggests that whenever variable changes can be organized, it is possible to uniquely relate a result to a variable. Although experiments are not run at manufacturing scale because they may jeopardize supply to the clinic, a manufacturing process can still benefit from DOE concepts by intentionally organizing the use of raw material lots [3] to avoid confounding variables. Just as for DOE in development, organizing variables so that raw material lot changes are not confounded will increase the ability to detect a change in a manufacturing setting and to understand the impact.

11.6 QbD FOR ORGANIZATIONS

Multivariate analysis not only applies to data processing and experimental design but also applies multiple ways to view a biopharmaceutical manufacturing process. This chapter so far has used two views, process flow and development timing. A third view deals with the dimension of corporate organization. In biopharmaceutical companies, there is a group of people who develop the manufacturing process while another group carry out the manufacturing. Not surprisingly, groups within a company have specific requirements to meet their differing goals. Development thrives on more information about the process. Manufacturing depends on operational simplicity and robustness. In some ways, these are competing needs: how can simplicity, cost containment, and increased understanding of raw materials coexist? Of course, these must coexist to achieve the highest quality products in a consistent manner. The nonobvious point is that by building in both aspects, the total process achieves maximal efficiency. The paradox of DOE holds for raw materials—more measurements really can mean less complexity.

11.6.1 Type I and II Risk

The organizational dimension described above boils down to a difference in information needs that can be quantitated. The result, leads to an understanding of how to organize analytical testing schemes applied to raw materials over the life of a product—a time-dependent QbD approach. The development scientist's view would be one willing to accept more type I error as defined in statistics. Specifically, a batch of raw material that would give acceptable outcomes in manufacturing process gets erroneously identified as unacceptable. The tolerance for type I error is higher because it would also come with more information about the raw material composition. A manufacturing scientist's perspective, however, might be the one willing to tolerate more type II errors and accept batches that differ from each other at the expense of having less information about a material. The tolerance for each error type will change over the development timescale for a product and will remain strongly influenced by the outcome of the manufacturing process. If a batch failure is observed in a manufacturing plant, for example, the desire for richer information content will increase and so also the probability of failure. Designing quality into the process for raw materials involves balancing the risks from type I and type II errors in a way that adapts with experience.

11.7 TESTS AVAILABLE

Small-scale use tests for critical, complex raw materials are widely implemented in the biopharmaceutical industry as a way to give definitive information about the probability that a raw material lot will perform acceptably in a manufacturing plant. A biological use test is considered the gold standard because it matches the real conditions as closely as possible. Because a use test consists of a scaled-down version of the actual conditions expected in the manufacturing plant, it provides assurance that no unforeseen variable or component in the material will adversely affect the process. Although much weight is given to results of a use test, it still suffers from run-to-run variability, takes approximately 2 weeks to conduct, is much more complex than more analytical raw material tests, and still does not provide understanding about factors that may underlie the performance. Analytical tests are preferable to use tests because they can generate information about underlying factors, but analytical tests also present risks. Since any analytical test will only be able to measure certain aspects about a sample, there is a risk of missing something important. Selection of an analytical strategy demands careful consideration to minimize the risk from unforeseen compounds entering the process as well as the risk from operational complexity.

As the experience with a raw material accumulates, these data factor into the probability that a false-negative or false-positive result will occur. Could an analytical testing scheme provide both types of information—intentionally robust and insensitive as well as intentionally "over" sensitive and informative with a definite strategy built in to transition between the types of information needed as the product development timeline moves forward? Bayes' theorem [4] can be used to understand the balance of these risks with time. Before exploring how Bayes' theorem can be applied, we need to understand more about the types of analytical tests that are available and the relative merits. Also, we will need some data to work with.

Analytical methods differ in many ways from each other, but one of the key differences has to do with their relative sensitivity and specificity. Bulk measurements like osmolality or pH are not capable of tracking the individual components that make up a complex raw material. A material could easily have the same pH or osmolality while having a different underlying composition than previous lots. These tests may not detect important composition changes and, therefore, be classified as allowing a high percentage of false negatives—a material would pass even if a minor component were not suitable.

NMR and MS, in contrast, provide unparalleled information about the chemical content in a mixture, but may be so sensitive, that even unimportant differences between batches of raw material could be detected. These tests would be classified as having a high type I error rate, but also high information content. Table 11.1 provides general classification for selected analytical methods based on their perceived error type.

MS and NMR can have such complex signals that multivariate analytical methods are used to interpret the data. This involves obtaining signals for a set of samples that will be used to "train" the method. As new samples get tested, they can be incorporated into the training set to improve the robustness of the model. During the initial phase, it is more likely that such a multivariate method will fail simply because the method is sensitive to factors that are not important to the outcome of the manufacturing process. Building a manufacturing process that has a tolerance for the time and temporary uncertainties associated with development of such multivariate methods might be one valuable outcome of a generalized "QbD" for raw materials strategy.

TABLE 11.1. Examples of Analytical Methods That Could Be Considered to Provide High False-Positive or False-Negative Rates[a]

"Too Sensitive" High False Positives	"Too Insensitive" High False Negatives
NMR	Visual Observation
LC-MS	Compendia[b]
ICP-MS	Osmolality
LC-DAD	pH

[a]A false-positive result would discover a difference between raw material lots that do not have an impact on production; a false negative would mean materials would pass the analytical test then impact production.
[b]Tests described in one or more of the following compendia USP, EU, EP, or ACS, and so on.

NMR, mass spectroscopy, HPLC, and ICP-MS are not commonly considered for routine use in raw material analysis. Each method depends on the consistent operation of very delicate and sensitive instruments. Part of the example presented below involved developing each analytical tool to see if it was predictive, robust, and rugged enough for raw material applications. The strategy used to develop each method included data preprocessing and statistical model development and was based on current advances in methods used for metabolomic applications.

Each of the analytical tests evaluated had the potential to become a platform technique for routine analysis of complex mixtures of raw materials in water. The techniques adapted were mass spectroscopy, NMR, and various types of HPLC. In each case, the idea was the same: take a specific, repeatable technique and interpret the results with pattern recognition software. Unique data pretreatment was evaluated for each type of data, then standard MVA was applied to help interpret the results. Details about these analytical tools are available elsewhere [5–17] and will be discussed here only to illustrate how QbD for complex raw materials encompasses not only analytical methods and multivariate analysis, but more importantly, a proactive strategy that helps ensure that quality is designed into a manufacturing process even when the degree of future variation cannot be predicted.

11.7.1 Hydrolysate Example

To illustrate the concepts for QbD applied to raw materials, consider an example involving a natural component added to media in a commercial biologic manufacturing process. Out of trend results for the bioreactor coincided with a change in the lot of hydrolysate. Small-scale bioreactor experiments confirmed that lot-to-lot differences of this raw material accounted for the change. An investigation ensued with two objectives: to devise an analytical test that distinguishes acceptable and unacceptable raw material lots and to identify specific components in the natural product that might account for the change. Although this was an investigation and not proactive process design, this

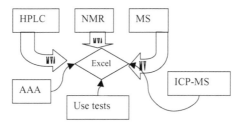

Figure 11.1. Information flow used to correlate analytical test results from different analytical methods with cell-culture performance. Data mining provides a retrospective example that will be used to gauge the probability that the performance from future raw material lots will adequately predict performance without the need for a use test.

example still informs what future actions might be needed to create a proactive QbD strategy for raw materials.

Figure 11.1 illustrates the strategy used during the investigation. Results from a use test were compared with scores from principal component analysis of HPLC-diode-array detection (LC-DAD), ^{1}H NMR, and LC-MS data. Results from ICP-MS and amino acid analysis were included without recourse to multivariate data processing.

LC-DAD, ^{1}H NMR, and LC-MS spectra each have noise as well as spectral regions with no relevant signals. To reduce the probability that these low-signal regions influence data interpretation, spectra were preprocessed to exclude uninformative regions and noise. First, for each method, raw signals were aligned to each other using correlation optimized warping as implemented in LineUp (Infometrix, Bothell, WA). Next, ANOVA was performed on all the channels from the dataset to highlight channels that vary by lot instead of replicating. Last, a threshold F-test value at 95% confidence was applied based on the degrees of freedom in the measurements. For LC-MS results, peak picking was performed prior to alignment using xcms operating in R [18]. To explore the results, the selected channels were compared using principal component analysis (PCA) or partial least squares (PLS) as implemented in MatLab (Mathworks, Natick, MA) with PLSToolbox (Eigenvector Research, WA) [19].

Methods were screened to see which could predict cell-culture results. For a result to be "significant" and retained, it had to have statistical significance on its own, independent of other measurement results. Factors that are significant only in combination with other factors were not included in subsequent data processing. Intuitively, this criterion helps ensure that only informative results are included in models that predict performance, but runs into trouble for methods that inherently combine information from several compounds at each measurement point—methods like ^{1}H NMR. For these methods, mixed factors in the form of scores for latent variables are used in place of individual component values.

To describe the process in detail, consider an example involving multiple hydrolysate lots tested using gradient reverse-phase chromatography with diode-array detection. At each retention time, a spectrum is recorded creating a three-dimensional data cube with retention time, wavelength, and intensity axes, for each injection (see Fig. 11.2). The method to compare results from different lots for all wavelengths and retention times is

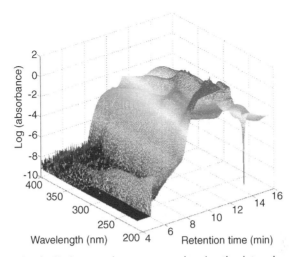

Figure 11.2. Example of a diode-array chromatogram showing the data cube generated for one injection of sample or control. (See the insert for color representation of this figure.)

well described in the literature [20]. The technique was applied here to assess many wavelength/retention time pairs at once.

After aligning retention time features at each wavelength for all the chromatograms with respect to each other, the entire chromatograms were compared by performing PCA. PCA organizes the data based on the amount of variation between chromatograms. The first principal component will contain the most prominent differences between chromatograms, followed by the next principal component, and so on. Figure 11.3 shows the scores for the first two principal components from the rp-HPLC-DAD analysis of six different hydrolysate lots. Lot 1 is separated from the other lots on the axis formed by principal component 2 (PC 2) that has 25% of the overall variation between

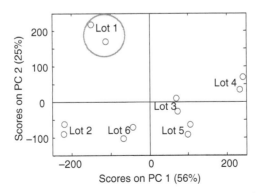

Figure 11.3. Scores for media additive lots from the first two principal components of the rp-HPLC diode-array results. The plot enables comparison of six hydrolysate lots, 16 min, 180 wavelength channels in one graph. Lot 1, which gave distinct performance in a cell-culture experiment is separated from the other lots along the axis for PC 2.

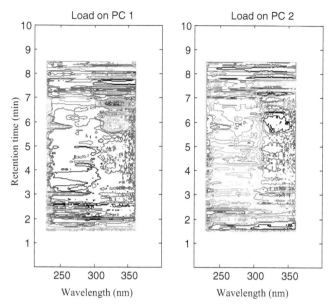

Figure 11.4. Principal component loadings that correspond to PC 1 and PC 2 in Fig. 11.3. PC 2 has an intensity pattern that suggests media additive Lot 1 (from Fig. 11.3) has more material absorbing at wavelengths greater than 300 nm and less material from 7.5 to 8.5 min. (Blue: less intense; red: more intense.) Not many differences are observed in the PC 1 loadings that reflect the sample scores plot. (See the insert for color representation of this figure.)

chromatograms. The pattern for PC 2 where Lot 1 is distinct from the other lots may be significant because the use test indicated Lot 1 was the only lot that performed well for cell culture. Next, the underlying signals that account for the distinction of Lot 1 will be examined.

Figure 11.4 shows the loadings for the first two principal components. The loadings reveal what spectral and chromatographic characteristics are responsible for the sample scores shown in Fig. 11.3. Because only PC 2 followed the pattern of interest to cell culture, most attention will focus on understanding the loading for PC 2. The images in Fig. 11.4 show color contours distributed by retention time and wavelength. Lot 1 has more intense signals (more red color in the image) at wavelengths above 300 nm between 2 and 7.5 min. UV-Vis absorbance spectrum of these lots confirms that Lot 1 has more relative absorbance in this wavelength region than the other lots (data not shown). The PC 2 loadings in Fig. 11.4 also reveal that Lot 1 has less material eluting from 7.5 to 8.5 min that absorbs 300–350 nm. Again, it is easy to confirm this observation by overlaying chromatograms in the 300–350 nm wavelength range (Fig. 11.5). Lot 1 has less material eluting between 7.5 and 8.5 min.

The benefit of MVA analysis applied to compare these chromatographic profiles is obvious from this small example. Without this MVA map, about 50 chromatograms would need to be compared, 12 chromatograms at approximately 5 wavelengths—too many to plot on one page. Using MVA, entire data cubes can be compared and interpreted

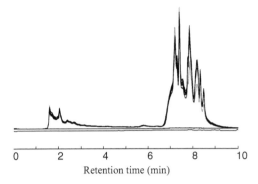

Retention time (min)

Figure 11.5. Overlay of reverse-phase chromatograms for the six lots of hydrolysate. Wave-lengths included are from 300 to 350 nm. This figure illustrates that results in Figs 11.3 and 11.4 are consistent with more traditional ways of summarizing analytical results. (Lot 1, gray; Lots 2–6, black).

all at once. Confirmation can then be targeted based on the results by plotting specific wavelength/retention time segments, as shown in Fig. 11.5.

Similar types of data processing were performed for ^{1}H NMR, LC-MS, and other HPLC separations. Results from each test were then correlated to cell- culture performance (Table 11.2). Not all analytical results correlate with cell culture performance. Analytical tests that do correlate to cell-culture performance with squared correlations above 0.25 are highlighted in green. Even after a test correlated to the use test when 6 hydrolysate lots were tested (like the HPLC example above), the correlation coefficient could drop when more lots are tested. For example, aspartate has a correlation coefficient of 0.5 when the first 6 lots are included in the analysis. When the number of lots increased

TABLE 11.2. Square Correlation Coefficient (R^2) and Sign of the Correlation Between Analytical Test Results and Cell-Culture Performance Data as a Function of the Number of Lots Tested

Method	Factor	R^2			Sign
		$N=6$	$N=11$	$N=22$	
AAA	Asp	0.5	0.4	0.0	+
	Ser	0.1	0.1	0.0	+
	Gln	0.0	0.3	0.0	+
	Trp	0.6	0.0	0.0	−
	Glu	0.0	0.1	0.1	+
	Met	0.0	0.3	0.0	+
	Thr	0.0	0.1	0.0	+
LC-Fluor	Unidentified Peak	0.6	0.1	0.0	+
ICP-MS	Zn	0.7	0.8	0.5	+
NMR	LV1 score	0.6	0.5		+
	LV2 score	0.8	0.3		+
LC-MS	m/z 279.17 intensity	0.5	0.4	0.3	−

to 22, the correlation coefficient dropped below 0.0. Other results, such as those for methionine, correlate strongly when 11 lots are included, but when 6 or 22 lots are included, the correlation is less than 0.0. Correlation coefficients for methionine and aspartate may be prone to type I errors based on the way the result fluctuates.

Four analytical factors have consistently high correlations with cell performance: Zn, the score from a PLS model based on ^{1}H NMR spectra, and the intensity of an LC-MS peak with m/z 279.17. Further discussion will focus on these three tests.

A QbD strategy for raw materials needs a prospective element that can allow reassessment of the test strategy with time. In the hydrolysate example, results from the use test influenced which analytical tests were conducted. Tests that showed a strong correlation to the use test were performed on additional lots while tests that did not correlate were dropped from future analysis. Understanding the benefits and costs of an evolving model that gets updated, as more data are available, will help move this example from the realm of a retrospective data mining into a forward looking QbD tool.

Figure 11.6 illustrates how the analytical testing strategy was refined over the course of this hydrolysate investigation. The left side of Fig. 11.6 shows that several tests were done early on, including some that did not correlate to performance of the cells, like AAA. Periodically, results from the analytical test were compared to the cell-culture outcome to look for correlations with parameters of interest. As new data became available, the conclusions were updated.

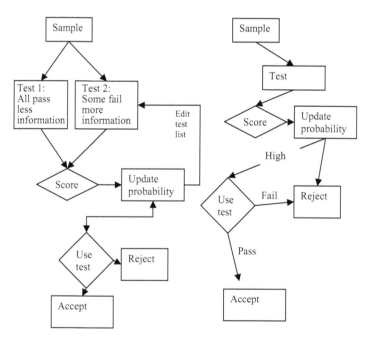

Figure 11.6. Flow diagram for the analysis of hydrolysates. The diagram to the left shows the investigation phase, the diagram on the right shows the final form.

The overall analysis depended on the sensitivity of the method and the cell-culture performance results. If there was no issue with the cell-culture performance, then an insensitive test would be acceptable. Every sample would pass the test with the benefit of no operational difficulty. In the hydrolysate example, the appearance of the powder would be an example of a test that was insensitive, having all batches pass. All the batches looked the same by eye and the test was easy to perform. Without data to the contrary, such a test might seem adequate. AAA, however, would be too sensitive, turning up differences that did not matter and spawning investigations while only providing some additional information about the hydrolysate. Once a batch failed during cell culture in manufacturing, however, the added information from AAA became important, but it was not possible to know how much weight should be given to the results. Taking this strategy to the extreme, a method that simultaneously measures everything, distinguishes raw material lots based on performance in the cell culture, and is easy to run would be ideal.

In the hydrolysate example, the use test from small-scale bioreactors was 100% predictive of the large-scale bioreactor. Based on the correlations in Table 11.2, what error would result if some combination of analytical tests replaced the use test? In this example, the answer was not clear. However, a threshold Zn level could be set that was sufficiently far from the prediction error for the method that some hydrolysate lots could be rejected based on the analytical result. However, lots with more Zn than the threshold value might still fail in production, despite the experience with the analytical test because other variables in the hydrolysate, the LV from NMR and m/z 279 (Table 11.2), for example, are also important. Without some ongoing estimate of these factors, a small risk of failure still exists. For this reason, a use test was still recommended on hydrolysate lots that exceeded the threshold level for Zn. The right hand side of Fig. 11.6 shows the testing strategy adopted following the investigation.

Ideally, Fig. 11.6 might consist of two analytical tests in place of the use test where the second test captures "unknown" compounds in place of the use test. Would NMR be an effective option? NMR is unparalleled for compound identification and has been applied extensively to complex solution analysis for metabolomic and food applications such as comparing urine, plasma, and beer samples [6,21,22]. In addition, NMR was recently described for the investigation of soy hydrolysates composition as it pertains to use in mammalian cell culture [2]. Either 1-D or 2-D NMR can be used for fingerprint analysis of complex mixtures. These techniques were applied in our hydrolysate example.

Figure 11.7 shows a typical ^{1}H NMR and HSQC spectra for a plant hydrolysate. Each chemical shift in the 1-D spectrum consists of a linear combination of signals arising from multiple compounds. For example, the methyl peak between 0.9 and 1.0 ppm (green arrow) consists of at least five different chemical moieties that could be from several compounds. To interpret these overlapping signals, 2-D NMR spectra like HSQC can be used.

Not all the signals in the HSQC spectra in Fig. 11.7 have a significant impact on cells. Highlighting only these "important" compounds can reduce the number of compounds that need to be identified and improve the NMR-based predictions. The first step was to reduce the number of peaks in the ^{1}H NMR. This was done using PLS comparing the spectra and cell-culture results from several hydrolysate lots. The number of peaks that trend with cell growth in the 1-D spectra decreased from hundreds in the raw spectra to 16

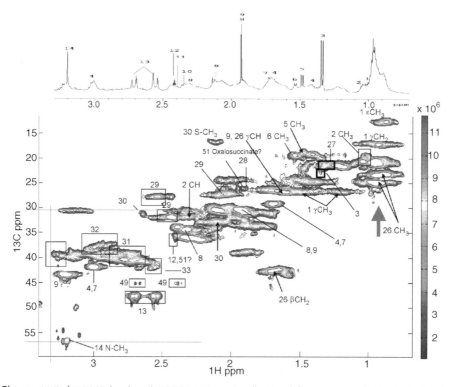

Figure 11.7. ¹H NMR (top) and HQSC NMR spectra (bottom) for a representative hydrolysate sample. Numbers indicate distinct functional groups or compounds that account for the signal. (See the insert for color representation of this figure.)

(Fig. 11.8, bottom). Next, PCA was used to identify peaks in the HSQC spectra that vary from lot to lot (Fig. 11.8, top). Peaks in the 1-D PLS correspond to a set of peaks in the 2-D spectra. Some of these peaks were identified while others could be prioritized for identification based on the correspondence between the 1-D and 2-D NMR results.

The correlation between ¹H NMR PLS scores and cell-culture performance is at least 0.5 (Table 11.2), rivaling the correlation of Zn. The types of compounds measured by ICP-MS and NMR are different; therefore, combining results from these techniques would increase the variables used to predict performance of a raw material. Combining scores from ¹H NMR and Zn may improve the predictive ability of the method and eliminate the need for routine use testing.

Table 11.2 points to three correlations. The last of these is an ion detected by LC-MS. The ion with m/z 279.17 has a consistent negative correlation with cell growth. When combined with the Zn concentration determined by ICP-MS, more variation is explained in a multilinear regression model. Likewise, by combining NMR and Zn results, the probability that the analytical testing would match that of the use test is increased. Efforts

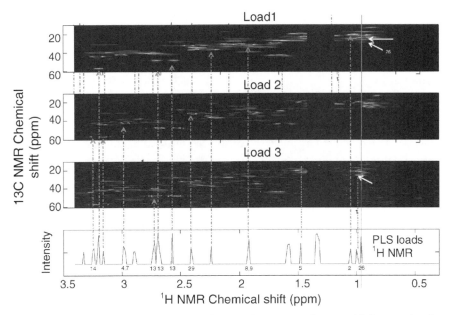

Figure 11.8. Comparison of three loadings from multiway PCA of HSQC with ¹H NMR loadings from a PLS analysis. Peaks that contribute significantly to the PLS analysis of ¹H NMR can be identified in the HSQC NMR loads. Identified compounds: 2, valine; 4, cadaverine; 5, free alanine; 7, lysine; 8, N-acetyl glutamic acid; 9, arginine; 13, citrate; 14, choline; 26, leucine; 29, acetyl phosphate. (See the insert for color representation of this figure.)

are underway to recast these types of data into a Bayes system to understand better the true risk to future manufacturing.

11.8 CONCLUSIONS AND FUTURE PROSPECTS

The complexity of biological drug manufacturing increases the need to reconsider raw material acceptance strategies in the light of QbD. An updated testing and release strategy for raw materials in manufacturing could result in a more robust manufacturing process. It should include a process to periodically assess and remove analytical tests that do not add value, maximizing both information and simplicity.

Current practices for raw material analysis demand ID testing on all incoming raw materials, but too often give no information on suitability of use with respect to performance. This chapter described a number of upcoming analytical techniques including HPLC fingerprinting, NMR analysis, ICP-MS, and LC-MS and described experience managing these methods and results as part of a retrospective investigation. The challenge, if raw material QbD is to be a truly meaningful phrase, will be in translating retrospective knowledge into a forward looking, adaptable strategy that can support continued learning about the relationship between raw material variation and manufacturing consistency.

Last, the examples described in this chapter point to limitations in the tools used to fuse data from several tests each with noise and to project these guesses into the future in some rational way. Estimation of process risk for unchartered variables—raw materials that will be received in the future—will be evaluated using more powerful statistical tools such as Bayes' theorem in the future. Bayes' theorem based calculations result in more stable and accurate estimates of errors than simple frequency-based results used in this chapter [23]. By combining types of analytical tests with different statistical and chemical sensitivities with a Baysian analysis framework, the prospective design space concept mapped out by QbD can be mapped onto future conditions of raw materials with more predictable risks and, therefore, an ability to highlight shortcomings that may creep into a control strategy as time passes.

ACKNOWLEDGMENTS

Dr. Julie Yu Wei, Biogen Idec, contributed Figure 11.7, experimental results and many helpful discussions.

REFERENCES

[1] ASTM International. ASTM Standard Practice E 2474. ASTM International, 2008 http://www.astm.org/Standards/E2474.htm.

[2] Luo Y, Chen GX. *Biotechnol Bioeng* 2007;97:1654–1659.

[3] Jorgensen K, Naes T. *J Chemometrics* 2004;18:45–52.

[4] Finn V, Jensen TDN. *Bayesian Networks and Decision Graphs*. 2nd ed. New York: Springer; 2007.

[5] Nord LI, Vaag P, Duus JO. *Anal Chem* 2004;76:4790–4798.

[6] Duarte IF, Barros A, Almeida C, Spraul M, Gil AM. *J Agr Food Chem* 2004;52:1031–1038.

[7] Araujo AS, da Rocha LL, Tomazela DM, Sawaya A, Almeida RR, Catharino RR, Eberlin MN. *Analyst,* 2005;130:884–889.

[8] Rezzi S, Axelson DE, Heberger K, Reniero F, Mariani C, Guillou C. *Analytica Chimica Acta* 2005;552:13–24.

[9] Forshed J, Idborg H, Jacobsson SP. *Chemometr Intell Lab Syst* 2007;85:102–109.

[10] van Nederkassel AM, Xu CJ, Lancelin P, Sarraf M, MacKenzie DA, Walton NJ, Bensaid F, Lees M, Martin GJ, Desmurs JR, Massart DL, Smeyers-Verbeke J Vander Heyden Y. *J Chromatogr A* 2006;1120:291–298.

[11] Debeljak Z, Srecnik G, Madic T, Petrovic M, Knezevic N, Medic-Saric M. *J Chromatogr A* 2005;1062:79–86.

[12] Lau AJ, Seo BH, Woo SO, Koh HL. *J Chromatogr A* 1057;2004:141–149.

[13] Gong F, Wang BT, Liang YZ, Chau FT Fung YS. *Analytica Chimica Acta* 2006;572: 265–271.

[14] Andersson FO, Kaiser R, Jacobsson SP. *J Pharm Biomed Anal* 2004;34:531–541.

[15] Christensen JH, Hansen AB, Karlson U, Mortensen J, Andersen O. *J Chromatogr A* 2005; 1090:133–145.

[16] Gong F, Wang BT, Chau FT, Liang YZ. *Analyt Lett* 2005;38:2475–2492.

[17] Gong F, Liang YZ, Fung YS, Chau FT. *J Chromatogr A* 2004;1029:173–183.

[18] Smith CA, Want EJ, O'Maille G, Abagyan R Siuzdak G. *Anal Chem* 2006;78:779–787.

[19] Watson NE, VanWingerden MM, Pierce KM, Wright BW, Synovec RE. *J Chromatogr A* 2006;1129:111–118.

[20] Gabrielsson J, Trygg J. *Crit Rev Anal Chem* 2006;36:243–255.

[21] Moreno A, Arus C. *NMR Biomed* 1996;9:33–45.

[22] Lenz EM, Bright J, Knight R, Wilson ID, Major H. *J Pharm Biomed Anal* 2004;35: 599–608.

[23] Gelman A, Carlin J, Stern H, Rubin D. *Bayesian Data Analysis*. New York: Chapman and Hall; 2003.

12

PAT TOOLS FOR BIOLOGICS: CONSIDERATIONS AND CHALLENGES

Michael Molony and Cenk Undey

12.1 INTRODUCTION

Process analytical technology (PAT) is the latest name given to a long-standing practice of engineers and scientists applying analytical tools to manufacturing processes to obtain new information, not gained from conventional analyses. Unlike other biologics industry-specific disciplines, this practice crosses the boundaries of many different types of manufacturing processes with the simple goal of increasing process understanding to the point of allowing some level of control. The desired output is to decrease the variability of the process resulting in a more consistent process with more predictable product quality. Several tool sets are at the disposal of the process analytical chemist and engineer. These include traditional laboratory analytical techniques applied online or in-line, automated laboratory control systems and software, multivariate data analysis techniques, and an array of next-generation analytical instrumentation and equipment specifically designed for process monitoring and control.

In biologics manufacturing, much of the current work involving in-line analytics emanates from the spectroscopic array, including Mid-IR, NIR, UV, fluorescence, Raman, and FTIR as well as the more commonly used simple control of pH, temperature,

Quality by Design for Biopharmaceuticals, Edited by A. S. Rathore and R. Mhatre
Copyright © 2009 John Wiley & Sons, Inc.

and gases. These analytical devices are being combined with statistical techniques made available by the discipline of chemometrics to gain new insights and better control of these complex biological processes. Other tried and true techniques are also increasingly used to gain process understanding. An assortment of chromatographic techniques such as gas and liquid chromatography are being coupled with a wide array of available detectors such as UV, fluorescence, refractive index, light scatter, charged aerosol detection, and mass spectrometry. Other nonsolid phase separation techniques are seeing increasing use due to the rapidity of their measurements such as flow injection analysis. Off-line spectroscopic techniques such as NMR, while still extremely useful, may benefit from the coupling of these devices to some of the latest generation of sterile sampling devices in combination with HPLC.

Data manipulation techniques such as signal preprocessing are frequently necessary to reduce background noise and optimize spectroscopic data for use in correlation building. The "scrubbed" data are then combined with other critical process parameter data (mechanical, chemical, and biochemical) collected by modern control systems such as dissolved oxygen, pH, sparge rate, nutrient attributes such as carbon source, vitamins, and certain beneficial trace elements. Often, detrimental metabolic indicators such as ammonia or acetate levels may be included in the data set to determine if buildup of the undesirable by-products should be more tightly regulated. The combined data set is then analyzed by using multivariate techniques such as principal components analysis (PCA), discriminant and cluster analysis, and partial least squares (PLS) analysis to understand the interdependencies of critical process parameters and potentially their effects on critical quality attributes. Multivariate exploration with correlations and cluster analysis enables looking at many variables at the same time. It is simply a set of statistical tools to look at continuous variables when they are considered as responses with no factors or independent variables. Used within the framework of continuous improvement, process analytical tools and multivariate techniques fit in well with the process improvement principles of six sigma and lean manufacturing and are enabling process scientists and engineers to work toward better defined and better controlled processes.

Relatively new fields such as surface plasmon resonance (SPR), microcapillary and nanofluidic arrays, and neutron reflectometry hold the potential to push the boundaries of process understanding further. Even tools not necessarily designed for bioprocess monitoring, such as hydrogen sensors or silicon dioxide monitors, are being applied to biologics processes and may lead to deeper understanding of biological processes.

Clearly, with many of these techniques, the ability to deconvolute the data and present them in a manner that is easily understandable in a manufacturing environment is critical to using these technologies. Integration of computerized laboratory systems is one of the components necessary to achieve real-time model building and automated process control systems. Supervisory Control and Data Acquisition (SCADA) systems, programmable logic controllers (PLCs), laboratory information management system, data warehouses, and adaptive multivariate analysis process tools are critical to achieving the type of control found in other industries such as the petrochemical industry or the food processing industry.

There are three areas in biologics process development and manufacturing that may benefit from these techniques. First, upstream processes (cell culture and fermentation) are demonstrating greater understanding of the nuances of media development as well as criticality of controlling the timing of process events and parameters (and materials) as a means of controlling process variability. Much work has been done on optimizing seed train transfer times using in-line and online techniques to more tightly control biomass. Second, downstream purification processes (such as chromatography and filtration) are coupling online FIA, HPLC, and chemometrics to explore what applications are conducive to relevant time decision making. And finally, formulation development is using compounding and fill/finish activities, including dosage delivery systems as well as dosage form processes such as freeze-drying, lyophilization, and crystallization, for exploring PAT applications to help control product quality attributes such as cake structure and uniformity, moisture, and other such attributes as excipient concentrations.

There are many challenges that exist during the course of development of these tools that must be dealt with in the framework of implementing best manufacturing practices. Most organizations discover that some of the tools are appropriate only in a process development environment. Typically, tools will be used to gain further process understanding and thereby obviating the need to put an equivalent tool in a manufacturing environment. Enough information about the manufacturing process will have been gained during the process development phase to allow a simple tactical or procedural solution to take the place of the tool. A certain subset of these tools may be used as platforms for cross-validation of a tool that is more conducive to a manufacturing environment.

As with any biopharmaceutical PAT applications placed in a manufacturing environment, the key to implementation is simplicity and reproducibility. The criteria that are of use in a manufacturing environment must be discussed and agreed upon with parties involved in the process development, automation engineering, and manufacturing functions. Additional negotiations will be required with personnel from quality assurance, quality control, and regulatory affairs in terms of how the technology adds value to the process and how to position the technique for regulatory inspections. Typically, this will involve presenting the process monitoring data to regulatory agencies within the framework of Quality by Design and protocol-driven exercises such as comparability. In the past, this type of data would have been deemed for information only and not presented as part of a regulatory filing. The paradigm has shifted and this type of data form the crux of the argument to move closer to design space concepts and allow the freedom to operate within a given range instead of at fixed set points. Further discussion can be found in Chapter 13.

The end goal of any of these technologies is to control the process in either a manual or automated fashion. Changes to manufacturing batch records may typically involve a migration from a process control point being made on a volume or time-point basis to decisions being made through calculated or empirically derived analytical results. Process control will then take the form of a trained operator manual switching of a valve or automated switching through a SCADA system in which the mechanism is controlled via a signal when the safety margin (upper or lower warning limit) for critical process parameter has been reached.

12.2   CELL CULTURE AND FERMENTATION PAT TOOLS

The richest and most rewarding environments for PAT exploration are in the "upstream process" disciplines of mammalian cell culture and fermentation. The practices of cell culture and fermentation have long been described colloquially as "black box" because of the less well-understood internal functions that take place when combining living organisms with plastic biobags, glass, and stainless steel. Several areas are being explored by process development scientists and engineers with current process analytical tools.

12.2.1   Miniaturization as Process Understanding Tool in Development

Microbial and mammalian expression process design space is being explored with the use of microbioreactor arrays in various formats and forms of automation. Arrays can range from fully automated robotic reactor arrays with full control logic to simple 96 well-based systems in which multivariate experiments may be performed rapidly. These systems may contain probes or sensors fabricated into the design to monitor various attributes such as pH, optical density (OD), and dissolved oxygen. Some of the more complex systems may rely on robotic workstations with adaptations to the robotic arm movements that include plate lifters that tilt cell culture plates to allow complete removal of media. Plates are transferred to 37°C CO_2 incubators and monitored and fed automatically with feedback loops that adjust for pH and feed additions by liquid dispensing systems. The entire station is enclosed within laminar airflow housing with automated UV decontamination. Other array-based technologies include SimCell™ miniaturized bioreactors that mimic the conventional controlled and monitored bench-top systems in a 150–1000 µL volume. The system uses microfluidics, permeable films, membranes, and advanced optical measurement systems to generate scalable processes. Core applications are for media development, process optimization, platform process creation, and clone selection [1, 57].

12.2.2   In-Line Analytics

In addition to conventional in-line probes for pH, dissolved oxygen, temperature, and carbon dioxide, other in-line probes are also used in fermentation and cell culture including optical density, capacitance, Raman, near-infrared, Fourier transformed infrared, fluorescence, pulsed terahertz spectroscopy (PTS), optical biosensors, *in situ* microscopy, SPR, and reflectometric interference spectroscopy (RIF). Early attempts at acoustic monitoring of bioprocesses have had little success. Many of these techniques are capable of measuring similar critical process parameters such as carbon source, pyruvate, lactate, and other metabolic by-products such as ammonia. Interfering spectral parameters, such as light scattering, path length variations, and random noise from variable sample matrix properties or instrumental effects, call for mathematical corrections. To reduce, eliminate, or standardize their impact on the spectra, data pretreatments prior to multivariate modeling are often necessary. Data generated by various in-line probes are typically preprocessed to reduce these types of noise due to bioreactor conditions associated with stirring and aeration (Fig. 12.1a).

The "smoothed" data (Fig. 12.1b) are then correlated with off-line data and using partial least squares models to build calibration and validation data sets, so that subsequent probe analyses will quantify the analyte of interest from the calibration set. When introducing this technology into regulated manufacturing areas, validation of the

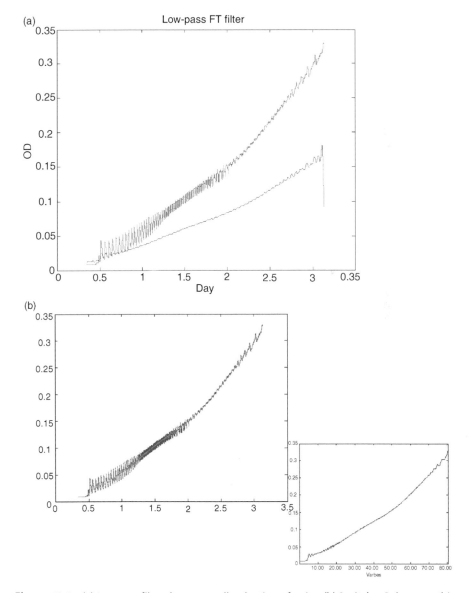

Figure 12.1. (a) Low pass filter shows overall reduction of noise. (b) Savitzky–Golay smoothing shows similar noise reduction but requires the algorithm to be run when evaluation each spectra.

off-line methodology is generally required prior to validation of the in-line technology. Sometimes, differences in the measuring technology off-line and in-line will not permit direct comparison. Model revalidation is necessary when changes to the processes occur that could influence the measurement device or the analyte of interest.

The OD probe has been used successfully to monitor biomass in fermentation reactors and attempts have been made to use it to measure biomass in cell culture but with significant obstacles. OD probes are useful in determining biomass in fermenters because of the high viable cell density and brief duration of most fermentation processes. Wu et al. [19] have tested several optical density probes and determined that a transmittance probe is the type best suited for determining total cell density (TCD) in mammalian cell culture. Since many mammalian cell culture processes run to lower than 70% cell viability, interfering absorbance from dead cells and cell debris may overestimate the actual viable cell density. In a head-to-head comparison of an optical density probe and a capacitance probe, Schmid and Zacher [20] determined that at lower cell densities, turbidity interferes with the ability to measure biomass accurately. For cell culture processes, optical density probes are better suited for applications where viable cell densities are maintained at high levels, such as during a seed train expansion step. If properly implemented, they should be able to provide timing and volume of seed reactors to subsequent seed bioreactor or to the production reactors. If the OD reading is slightly out of tolerance, adjustments can be made to the volume delivered to maintain the consistency of the total cell density transferred. To make the technique usable in cell culture, regression coefficients obtained from the calibration cure are used to predict TCD on the basis of OD readings.

Capacitance probes, or dielectric spectroscopy probes, are the other type of probes used to assess viability in mammalian cell culture. The benefit of a capacitance probe is that it measures capacitance of intact cell membranes, so only product-producing cells are measured. In addition, capacitance probes are less sensitive to dead cells, bubbles, and dissolved solids than optical density probes since they do not rely on light scatter. It has been clearly demonstrated that these in-line probes yield readings that tightly correlate with off-line viable cell densities by automated hemacytometer or other such devices (Fig. 12.2).

One concern with capacitance probes is their ability to give accurate readings with cell cultures that have different cell morphologies within the same culture. While the permittivity at a given frequency is linearly related to viable cell density (biovolume), some probes have been designed with parallel multifrequency measurements (dielectric spectroscopy) to access the information on different physiological and morphological cell phases. The very limited ionic permeability of the plasma membrane gives living cells specific dielectric properties. At high frequencies, above 10 MHz, the ion movement amplitude is limited and is not significantly affected by the presence of the cytoplasmic membrane. A cell suspension has nearly the same overall electric properties as the suspending medium. At lower frequencies, around the critical frequency (fc), the displacement amplitude is higher, ions accumulate at the membrane, and the cells behave as tiny capacitors. The global charge accumulation depends on the viable cell density. When measured at a fixed frequency close to the average fc, the permittivity is independent of cell size and state. At very low frequencies, when the cells are fully

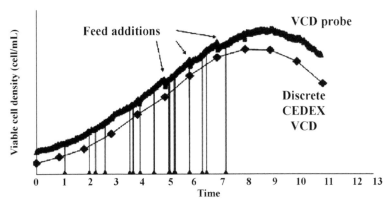

Figure 12.2. Real-time capacitance probe trace showing the correlation between off-line viable cell density measurements by hemacytometer (CEDEX). Note that the spikes in the viable cell density trace correlate with feed days during a mammalian cell culture fed-batch process.

polarized, there is no further capacitance increase related to charge accumulation. The increase in permittivity from high to low frequencies has a sigmoidal shape and is known as the β-dispersion [22].

The NIR probe is one of the quintessential PAT tools in the process scientist and engineer's toolbox. With the recent advances in fiber optics and the broad availability of chemometrics software, it is now common practice to perform *in situ* NIR analysis by submerging the probe into the reactor. It is based on the absorption of electromagnetic energy in the region of 700–2500 nm, caused by the overtone and combination bands of the fundamental bending and stretching vibrations as seen in the mid-IR region [23]. These NIR absorptions are generally 10–100 times weaker than the fundamental bands of mid-IR [25], but this is in fact an advantage, as it enables the direct analysis of samples without any sample preparation. Therefore, it is ideally suited for real-time measurements, and although the spectra appear broad and overlapping, much information can be elucidated from them by using sophisticated chemometric techniques [24]. From first principles, therefore, NIRS is perhaps more suited for application *in situ* than any competing technology [56].

In situ NIR has been implemented successfully in both microbial fermentations and cell cultures. As with most of these spectroscopic techniques, chemometrics is critical to building predictive, quantitative models. Parameters fall into two main categories: pretreatments and diagnostics. Pretreatment of data usually takes the form of derivatization, smoothing, and path-length treatments. Diagnostics include the tools necessary for demonstrating or quantifying the success of a calibration. Typical calibrations involve an "add back" model, as described by Yeung et al. [4], where the analyte(s) of interest is (are) added to the matrix in which they would be found. Validation runs use "unknown" samples and spiking of unknowns with knowns and quantifying them against the calibration model and correlating the values to an orthogonal off-line analysis.

For fermentation processes, NIR has been extensively used in antibiotic production, recovery of a yeast alcohol dehydrogenase (ADH) from an unclarified yeast cell homogenate as well as biomass in *Escherichia coli* fermentations [1–7]. Arnold et al. [7] have even developed a temporally segmented model in which they built predictive NIR models for the early, middle, and end phases of a bioprocess. Since the nature of the matrix they were working with changed during the course of the fermentation, they were able to successfully create predictive models for each phase of the process. The chemometric technique of partial least squares was used to build the models. The models were built on wavelength regions, rather than on distinct wavelengths, in a more complex matrix such as fermentation broths [5]. Data segmentation may be a useful technique when dealing with complex matrices that change physiologic properties over the duration of the process. Data segmentation is a powerful diagnostic that may spot differences between runs but will not necessarily tell you why they are different.

NIR has been used to quantify the amounts of metabolic by-products in cell culture. There are numerous examples reported in literature on the NIR probes' ability to quantify key metabolites such as carbon source (glucose), lactate, and ammonia, as shown in Fig. 12.3 [12–18].

Card et al. [17] reported using a first derivative with a Norris smoothing and mean centering lead to quantify a number of cell culture metabolites and variable including glucose, glutamine, ammonia, cell density, and cell viability. In addition, they showed that lactate required a slightly different treatment and showed the best results with a slightly more aggressive smoothing and a combination of variance scaling and mean centering. Total cell density calculations were optimized with no spectral pretreatment except for a mild Savitzky–Golay smoothing and mean centering. They found most of the useful cell density signal was contained in baseline information, either offset or slope, or a combination of both [17]. They used this information to control the fed batch process by altering the feed components.

FTIR with attenuated total reflectance, as implemented in bioprocesses, is able to monitor many of the same attributes as NIR. FTIR has been successfully implemented in both microbial and mammalian bioprocesses. In some instances for online usage, a thermoelectric cooled mercury cadmium telluride (MCT) detector is used. Analytes that have been reported to be measured in microbial fermentation include glucose, sucrose, ethanol, citrate, lactate, glycerol, methanol, yeast extract, ammonia chloride, ethyl acetate, and a variety of other components. Schustera, Mertens, and Gapes [28] report using FT mid-IR to monitor the acetone–butanol–ethanol (ABE) fermentation in the genus *Clostridium*. The ABE fermentation process is a technology that converts renewable resources into liquid fuels and basic chemicals. In cell culture, FTIR has been used to measure relative amounts of DNA, RNA, lipid, protein, and glycogen throughout the different growth phases of a culture [29].

Like other spectroscopic techniques, FTIR relies on chemometric techniques such as partial least squares to build quantitative models, but these models assume linearity of response and thus may introduce errors so significant as to make the spectroscopic technique unsuitable for implementation. Therefore, ensuring the use of the appropriate chemometric technique(s) to fit the process being modeled is imperative. For example, Franco, Perin, Mantovani, and Goicoechea [30] were able to accurately measure glucose, glucouronic, and gluconic acid by combining strategies of nonlinear partial least squares, a wavelength-selective genetic algorithm, and artificial neural networks.

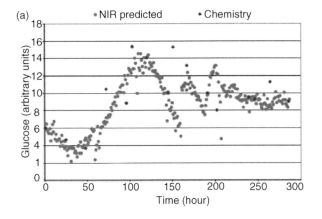

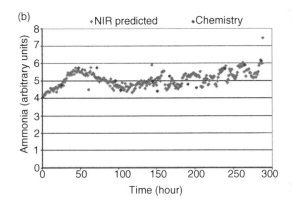

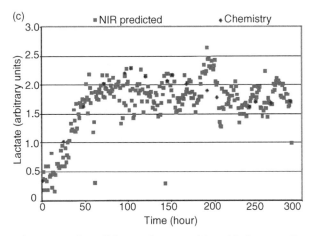

Figure 12.3. NIR plots comparing off-line analyte quantities with those predicted by calibration sets derived from partial least squares analysis of carbon source—glucose (a) and waste by-products— ammonia (b) and lactate (c).

Raman spectroscopy relies on inelastic scattering involving an energy transfer between incident light and illuminated light target molecules. Problems in applying Raman spectroscopy to bioprocesses lie in the strong fluorescence activity of many biological molecules that often overlay the Raman scattering bonds. In bioprocesses, it has been used to quantify the production of total intracellular carotenoid astaxanthin from yeast and alga. Lee et al. [10] have shown quantification of glucose, acetate, formate, lactate, and phenylalanine in *E. coli* fermenters.

Optical bio- and chemosensors are most commonly used fluorescent dyes immobilized on the tip of a fiber optic cable. Online measurement of various process parameters such as pH, CO_2, or O_2 can be quantified depending on the dye used. Because the tips are replaceable, this technology offers a broader range of applications than traditional sensors for pH and off-gas analysis. The change in fluorescence intensity of a metalorganic dye, caused by quenching of oxygen, is the measuring principle of fiber optic oxygen sensors. This technology is being taken to the next logical progression for monitoring bioprocesses by immobilizing enzymes, antibodies, oligonucleotides, and in some cases even whole cells selective for particular compounds. These affinity sensors for protein analysis during fermentation and downstream processing have included SPR and RIF. Most manufacturers of these two kinds of optical biosensors supply instrument-ready surfaces and applicable reagent chemistries for ligand immobilization [31]. The modes of detection are also expanding to include colorimetric, chemiluminescence, bioluminescence, and electrochemical methods to solve particular process problems.

Pulsed terahertz spectroscopy has become a preferable method for performing low-frequency spectroscopy over standard FIR (Fourier infrared) techniques. A pulsed terahertz experiment is similar to conventional pump-probe setups. A terahertz beam interrogates the properties of a sample ("pump") and is superimposed on a detector together with pulses of a second ultrafast beam ("probe"). Varying the time delay of the beams relative to one another enables the electric field of the terahertz wave to be reconstructed, and a Fourier transform finally yields the desired spectral information.

Engineers at some of the major optics companies are reportedly working at PTS modifications to allow determination of the sequence of intermediate tertiary structures (similar to time-resolved CD) in bioreactors and fermenters. An example is ECOPS (electrically controlled optical sampling) systems that use two ultrafast lasers that are phase stabilized to each other. A slight modulation of the length of one of the fiber oscillators via a piezo element serves to sweep the "probe" pulse through the terahertz pulse in a precisely controlled manner. PTS is a viable technique for future time-resolved FIR measurements of protein folding. Thus, it could become a very useful tool for bioprocess monitoring, since the correct folding of recombinant proteins is one of the most important factors in the biopharmaceutical industry [31].

Given the high utility of online probes, the most difficult question the bioprocess engineer has to answer is "Which and how many of these technologies are worth implementing in my process?" Although NIR is currently viewed as a panacea because it can measure what most of the other probe types can, it is not always the right tool for the job. For example, in an industrial process for viral production, an OD meter may best serve the purpose for measuring cell lysis by monitoring increasing turbidity. A question that bioengineers need to ask themselves is "Is there truly a need for continuous real time data or will a less frequent sampling rate suffice?" If the aim is to gain process understanding for metabolites, but conventional off-line analyses or a more frequent sampling rate, enabled by an autosampling system will meet the requirements, then a probe may not be the best tool of choice. It is reasonable to assume that unless the overall objective of the tool is to use the output to control the process such as optimizing nutrient feed times, evacuation or neutralization of deleterious by-products, or continuous monitoring of certain critical process parameters on which process control will be implemented, then an in-line solution may not be required. With in-line probes, there are a multitude of risks that should be assessed prior to implementation that include cleaning, validation, maintenance, training, process appropriate model construction, and, in the case of cell culture, risk of bioreactor contamination.

12.2.3   Online Analytics

Sterile Sampling Systems.   Online sterile sampling systems are the linchpin in successfully interfacing conventional off-line analytics with microbial fermenters and cell culture bioreactors. Sampling can be passive and rely on gas or fluid dynamics to deliver it to the analytical device or it can be active and use any number of pumping mechanisms such as peristaltic or syringe pumps to actively pull the sample from the tank and then push it into the analytical device. In fact, a group was formed to specifically address these problems in the late 1990s at the University of Washington Center for Process Analytical Chemistry called NeSSI™ (New Sensor and Sampling Initiative). It is a consortium of academia, instrument manufacturers, and industry users to help design, build, and test new sampling and test devices. It focuses on standardization and miniaturization of online sampling and analytics and has slowly infiltrated areas beyond its initial chemical and petrochemical industry roots to find applications in the automotive, food, and pharmaceutical and biopharmaceutical industries.

When online sampling devices are coupled to bioprocesses, a reasonable level of assurance for the prevention of microbial contamination must be provided to prevent potential contamination from either backflow, once the sample stream passes the reactor barrier or leaks at the port of entry. These devices are increasingly making their way into biopharmaceutical development lab fermenters and bioreactors where they are being coupled with existing analytical instrumentation to gain new insights into the process during the development phase of a molecule. A passive sampling device coupled to an ion chromatography HPLC with a pulsed amperometric detector was used by Larson et al. [26] to monitor amino acids and glucose during the course of a bioreactor run. A syringe pump-driven sampling device was coupled to a conventional HPLC to yield information such as amino acid consumption during the course of a reactor run [8] and even some product quality attributes with the addition of two-dimensional HPLC (Fig. 12.4).

Monoclonal antibody development, in particular, is conducive to the two dimensional HPLC approach from bioreactor streams, as the first dimension separation is usually an affinity separation such as protein A or protein G chromatography. This is then followed by a second dimension separation that monitors a particular quality attribute such as ion exchange, reverse phase, or size exclusion. With these tools, it is then easier

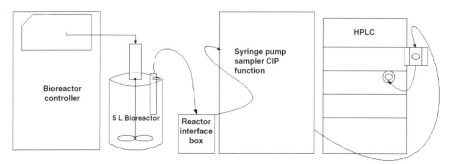

Figure 12.4. Development lab-scale online HPLC configuration for monitoring critical process parameters and some critical quality attributes.

for the process developer to understand how the bioreactor conditions, feed media content, and feed times affect these critical quality attributes. Design of experiments (DOEs) coupled with multivariate regression tools are typically used to determine the critical process parameter with the highest contribution to the desired (or undesired) critical quality attribute. In general, online sampling devices coupled to two-dimensional HPLC allows the bioengineers to develop a first dimensional separation that focuses on purification of the protein of interest from the feed stream and a second dimensional separation that focuses on the product attribute of interest. The same logic applies to process analytes of interest such as looking at trace elements and sugars by ion chromatography, although the in-line sample preparation becomes critical. In reality, the ability to deliver samples from bioreactors straight into empty HPLC vials allows a variety of derivatization chemistries to be performed by using the injector programming features present in most modern HPLC autosamplers. These samples may then be separated and detected by UV, fluorescence, or mass spectrometry. While appropriate for gaining process understanding in a development lab, the state of these systems, as they presently exist, is rarely implemented in commercial production bioreactors due to the complexity of operation, maintenance of the equipment, and impact on the manufacturing process if the instrument fails.

In development labs, the output data from these in-line sterile sampling systems coupled to traditional analytical devices are currently being input into PLCs through standard system connectivity languages such as OPC (open connectivity). The PLCs that receive the information control peristaltic pumps that are actuated when set points are reached. Based on the incoming data stream from the online analyzer, once a set point is reached for a given critical process parameter, the PLC activates the peristaltic pump to begin pumping feed media into the reactor, thus semiautomating a fed-batch process.

At this time, there are relatively few commercial systems available to perform these types of tasks in a GMP biologics manufacturing environment. At present, only a handful of equipment manufacturers make HPLCs suitable for the manufacturing floor, and there are even fewer types of sterile sampling systems necessary to perform automated sampling. As drug manufactures look for more "relevant time" process analytics related to process and product, it is expected that more commercial systems will become available for GMP use in near future.

Direct Instrument Interface with Sample Stream. The most common example of an instrument that connects directly to a bioreactor or fermenter is a gas chromatographic mass spectrometer. Progress has also been made in directly coupling bioanalyzers or flow injection analysis units to bioreactors. Each of these systems is capable of measuring off-gas analysis, such as CO_2 (waste aerobic cell cultures) and O_2 that is nutrient in aerobic cell cultures and used for calculations in oxygen uptake rate (OUR). There are generally two methods of determining OUR: the first is calculated (K_La and total air sparge) and the second is empirically determined (mass spectroscopy-based gas analysis). For the calculated, two assumptions are made that contribute to high process variability; the first is that K_La for clean water and the culture broth are the same. The second is the assumption of a pseudosteady-state where OUR is equal to OTR (oxygen transfer rate). Oxygen transfer contributes to process variability by affecting the overall cells' metabolic ability to convert carbon sources to energy in the form of ATP via the Krebs cycle. Since these gases are used to monitor the overall health of a bioreactor during the course of a multiple day run, it is worth exploring the option of empirically determining these values to more precisely control the process.

In addition, bioanalyzers are capable of generating other valuable information with respect to metabolites and waste products that may be used to control bioreactors. The instruments use two types of electrochemical analyzers: potentiometric and amperometric. Potentiometric sensors provide basic parametric measurements, such as pH, and measure various other metabolic waste products such as ammonium, sodium, potassium, and acetate. Amperometric sensors are the real heart of bioanalyzers and are configured to measure metabolites and nutrients such as glutamine, glutamate, glucose, and lactate, among others. In some instruments, a photometric sensor is used to detect compounds such as phosphates or glycerol. Keeping the correct balance of essential nutrients and metabolic waste products can be optimized with such systems that allow control feedback loops to adjust timing and amounts of nutrients delivered during the bioreactor runs. Without this continuous stream of feedback, the result is the fed-batch paradigm that usually leads to higher process variability and often higher product quality attribute variability.

12.3 PURIFICATION PAT TOOLS

12.3.1 Adaptive Control of Buffer Concentrate Dilution and Chromatography Skids

PLC-controlled blending and dilution skids are commercially available that perform precision in-line dilution of buffer concentrates often using the plant WFI system. Sensors are used to measure the dilution or blending and adaptively control the blending streams to arrive at the desired concentration or blend ratio. This level of precision allows tighter control of conductivity and pH during the course of separation thus allowing less noise in the UV trace. These technologies also offer the ability to minimize manufacturing plant footprint by reducing the need for large storage tanks.

Similar technology is used to control process chromatography skids for higher precision mixing during isocratic, step, and gradient elution with the additional benefit of in-line optics for control of fraction collection. These systems are designed to minimize allowable tolerances for buffer compositions at any given stage of a process by adaptively controlling mixing rates.

Natarajan and Purdom [40] designed a system to monitor the formation of a salt gradient used in an ion exchange process step by measuring the conductivity of the effluent solution. They demonstrated that the shape of the resultant gradient was nonlinear and resulted from valve dynamics and the nonlinearity of the mixing of solutions of different densities. They developed a feed-forward control strategy based on a mathematical description of the dependence of the effluent conductivity on the mixing ratio. Their PLC logic employed a quadratic fit model based on variation of effluent conductivity at various mix ratio values along with the desired conductivity at a given time to monitor the accuracy of the control. With this system they demonstrated an improvement in the accuracy of the gradient.

12.3.2 Online Monitoring of Downstream Processes

Although online HPLC used to control downstream processes is receiving more attention nowadays, the technology has been successfully employed in downstream bioprocess manufacturing since the early 1980s [39, 42]. It has been demonstrated to provide benefits such as increased operating efficiency, cycle time reduction, step coupling to produce semicontinuous operations, reducing the possibility of processing errors, eliminating manual handling of fractions and samples, reducing the opportunities for product contamination, minimizing variability, enabling feedback control of critical process parameters, and increasing throughput capabilities.

Cooley et al. [42] used online HPLC to reduce the variability of eluate collection caused by column loading, the purity of the starting material, the affinity of various components in the process stream for the stationary phase, and the column's operating parameters such as the generation of the gradient used to elute the column. This variability generally leads to conservative collection set points that may reduce product yield.

In Cooley's system, the online HPLC sends the product purity value, derived from the online HPLC data, to the distributed control system (DCS). The product purity value generated by the online HPLC is compared with the product purity set point in the DCS. If the product purity value is greater than or equal to the set point, the eluate is sent to the cooling tank. This allows a more consistent process output based on product quality decisions. The key to any such implementation is that the speed with which the analytical results are generated must be amenable to the timing of the process step. Because the process steps that the scientist and engineers were monitoring allowed "relevant time" feedback, they were able to leverage this strategy at a later point. They designed a more complex system to couple two process ion chromatography steps by the introduction of the pure fraction from the first column to be loaded, via a switching valve, onto a second column.

As is more often the case, other scientist and engineers tried similar techniques at development scale, but business decisions were made not to proceed forward to production scale. There are quite a few examples of implementing online HPLC in a development environment, which never proceeded to be implemented in the manufacturing process. Fahrner et al. [33] demonstrated online reverse-phase HPLC for separating aggregates from recombinant human insulin-like growth factor-I (IGF). Rathore et al. [43] have shown many instances in which separations were evaluated with respect to their ability to control processes. First, for controlling a protein refolding step, they looked into the feasibility of designing a control strategy using online monitoring that would allow refolding operations to end at a time determined by product quality parameters. The data suggested that while possible it might not be practical due to the "complexity" issues in a manufacturing environment. Rathore states concerns for using online methods versus the simplicity of a time-based approach. Concerns such as training operators to make decision with an online analytical method as well as redundancy for a fail-safe mechanism would be obstacles in implementing any such technology. In the same set of studies, they looked at controlling an end point analysis to the UF/DF step based on a product quality measure. In the final study of the set, they evaluated a control strategy for ion exchange fraction pooling based on desired product purity. They analyzed the samples by reverse-phase HPLC for product purity and demonstrated the ability to control product purity through such a method, while stating similar rationale for not implementing the strategy in a manufacturing environment.

In a separate body of work, Rathore et al. [11] demonstrated implementation of online size-exclusion HPLC to control fraction purity pooling from a process step that used a hydroxyapatite column separation. The system was set up to collect sample from a side stream.

They were able to show the feasibility of implementing the online analysis for facilitating a real-time decision making for pooling of chromatography column based on product quality attributes. Experiments were performed with low purity (65.5%), moderate purity (71.3%), and high purity (76.1%) and feed material. Column eluate fraction pooling was performed using preset criteria to halt the pooling process when the fraction purity reached 85%. They were able to achieve a pool purity within 1% variation despite a 10% variability in the purity of the feed material. A processing paradigm that Rathore points out is that by targeting consistent pool purity by shifting pooling criteria, the recovery across the chromatographic step varies. Rathore also points out the fact that regulatory authorities have historically viewed step yield as a measure of purification process consistency. If a company were to implement such a tool, a regulatory strategy would be needed in a filing, of any such process, to address this potential regulatory concern.

Lanan and McCue [38] demonstrated that online reverse-phase HPLC was useful to purify low molecular weight impurities from the product of interest during the course of an ion exchange step. Off-line studies showed that low molecular weight impurities eluted in early fraction of the ion exchange, so the goal was to monitor the low molecular weight impurity stream to determine when to begin the collection of the protein fraction of interest. In the online system at pilot scale, a stream splitter was connected to the outlet tubing of the ion exchange column. The tubing from the splitter was connected to a rheodyne valve with a fixed volume injection loop. The loop was switched in-line with the reverse-phase column at predetermined periodic intervals to perform the reverse-phase separation to separate the low molecular weight impurities. They were able to

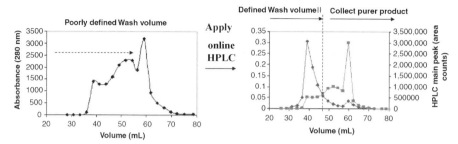

Figure 12.5. Online reverse-phase HPLC for ion exchange fraction pooling to monitor removal of low molecular weight impurity.

demonstrate that the ability to use the data from the reverse-phase separation to define the wash volume based on removal of low molecular weight impurity instead of a predetermined number of column volumes (CVs) used in the wash step (Fig. 12.5).

In a separate set of experiments, Lanan, Kiistala, and Parikh [38] demonstrated the ability to monitor a protein pegylation reaction to completion by size exclusion chromatography. A column switching valve was placed between a sampling probe outlet line from stainless steel pegylation reactor vessel and the analytical HPLC. A peristaltic pump was used to draw sample from the vessel into the HPLC tubing into the sample loop on the rheodyne injector valve. The sample loop was switched in-line at predetermined periodic intervals and a size-based separation was performed from the beginning of the pegylation reaction to completion. Visualization of the protein of interest, the monopeglyated species of the protein, and the multipegylated species of the protein can be achieved by plotting stacking the chromatograms from earlier sampling points on top of chromatograms from later sampling points (Fig. 12.6a).

By plotting all peak areas from the multi-PEG, mono-PEG, and the raw material PEG, Lanan, et al. were able to visualize a real-time assessment for the completion of reaction. With the batch of polyethylene glycol used in this experiment, pegylation of the protein was complete in approximately 10 h (Fig. 12.6b). This specific application of a PAT tool would allow the process decision to be made on the basis of an "end point" analysis, rather than a set time period. If raw material variability in such reactions were a concern, then having a window into the process might be considered advantageous, as it would lead to the same product quality regardless of the raw material variability.

In addition, flow injection analysis (FIA) is making a comeback as a PAT tool for purification. Almeida et al. [45] used FIA to demonstrate the ability to make *Fusarium solani pisi* cutinase assessments from an expanded bed absorption eluate using micro-encapsulation of *p*-nitrophenylbutyrate (*p*-NPB) in a micellar system. They were able to distinguish slight differences in yeast cultivation conditions during cutinase production that influenced the fermentation performance that affected the adsorption of cutinase during resin loading. They demonstrated a good correlation between the FIA system results and the off-line cutinase activity results. Putting this type of system in-line and using either principles of hydrodynamic chromatography or field flow fractionation may allow process engineers to assess purification fractions based on larger molecular weight heterogeneity without the addition of a stationary phase. This approach would also solve

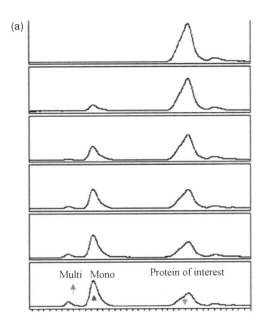

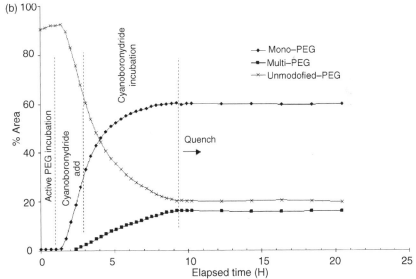

Figure 12.6. (a) Protein pegylation monitoring by online HPLC size exclusion chromatography. (b) Relative distribution and required reaction time for mono- and multipegylated species by online HPLC size exclusion chromatography.

some of the issues of putting a complex instrument in the manufacturing plant, as these systems are simpler by definition.

In summary, for companies looking to implement online analytics as a purification PAT tool, the ability to overcome corporate risk aversion associated with introducing a "complex system" into a production environment is likely still the single largest obstacle. Comfort level of the process developers—quality, regulatory, and manufacturing—must be vetted after the proof of concept work has been completed to gauge if the technology would be readily accepted into the manufacturing environment. Some scientists and engineers have suggested that training of manufacturing personnel to operate complex machinery may be an implementation barrier, but simple redirection of existing resources, by shifting the quality paradigm and putting QC analyst in the manufacturing plant, may be a more risk-appropriate solution. Regardless, until such perceptions are overcome, these types of tools will likely remain in the process development labs to increase purification process understanding.

12.4 FORMULATION PAT TOOLS

12.4.1 Drug Substance

Traditional bulk drug substance biopharmaceutical freeze–thaw processes entail large bottles or carboys placed in walk-in freezers. This leads to a poorly controlled freezing process experienced by the bulk drug substance. Conversely, thawing is traditionally performed in a temperature-controlled water bath, which may result in contamination if container integrity is compromised during or before thawing. Out of this need, commercially available units have arisen for controlling freeze–thaw cycles [52]. This is an area where PAT tools are being coupled with disposable technologies to allow more flexible and modular unit operations.

Critical process parameters for bulk drug substance include proper formulation of the correct combination of buffers, bulking agents, and polydispersants, and the controlled freezing and thawing of these solutions to minimize product degradation. Of primary concern to most biopharmaceutical manufacturers is the formation of soluble aggregates or precipitants during the freezing process. Aggregate formation and other chemical modification such as oxidation or deamidation can be triggered by changes in temperature, pH, ionic strength, and excipient concentration gradients formed during uncontrolled freezing and thawing.

Flexible disposable containers up to 100 L are commonly being used to store biopharmaceutical bulk drug substance. The benefits of such a container are that it can be pressed between heat exchange plates containing a circulating liquid to control both freezing and thawing processes. The large surface-to-volume ratio of such a system favors rapid thermal transfer. Monitoring of product and heat transfer fluid temperatures is critical to documenting process uniformity and reproducibility. The result has been described as a frozen "brick" that is advantageous for shipping and storage compared to frozen carboys or stainless steel vessels.

12.4.2 Drug Product: Lyophylate in Vials

PAT tools for formulation as well as fill/finish activities are well documented and are another discipline where spectroscopic techniques are frequently used to make relevant time and product quality checks to make process decisions. Near-infrared spectroscopy is used for moisture as well as cake uniformity and structure for solid dose biopharmaceuticals [44].

Lyophilization is usually performed to increase the shelf life of biopharmaceuticals susceptible to degradation in the presence of water. High cake porosity, low residual moisture, and an "elegant" presentation state are the most prominent quality criteria of lyophilized products. Traditionally, the moisture content of lyophilized products is determined by time-consuming methods, such as Karl Fischer titration, but newer methods such as vapor sensor technology are reducing the amount of labor required to obtain a moisture value.

NIR can be correlated with moisture by Karl Fisher or by the gravimetric measurement typically used for loss on drying (LOD) determinations. Due to the strong overtone bands for water at 1940 nm and 1450 nm, it makes it an ideal nondestructive method for quantification of moisture. Calibration sets are prepared by introducing small amounts of water onto the container closure walls of a sealed vial containing lyophilized cake, being careful not to let the water touch the cake. Upon introduction of the different volumes of water, the vials must be allowed to evaporate and equilibrate for at least 2 days in a position that does not allow the water droplet to travel down the wall of the vial and be introduced to the cake. Diffuse reflectance NIR calibration samples can be acquired and a multiple linear regression (MLR) model may be used to correlate the NIR data with an orthogonal method. Moisture is also a cause of cake "meltback" or collapse; thus, NIR spectroscopy may be implemented in an automated fashion to perform this nondestructive analysis on an "every vial" basis rather than a statistical sampling basis.

In addition to being a nondestructive moisture determination, NIR diffuse reflectance techniques can help quantify cake content and uniformity and quality of the reconstituted. Reich and coworkers [49, 50] reported the use of NIR spectroscopy to evaluate stress-induced structural changes of proteins and stabilization effects of sugars upon lyophilization, storage, and rehydration. Spectra of stressed and unstressed proteins revealed changes associated with the primary, secondary, and tertiary structure of the proteins. Sensitive amide I, II, and III bands and the water absorption band may be used for the assessment of protein structural changes and aggregation.

As discussed by Pikal et al. [47], manomeric temperature measurement (MTM) is another PAT tool being explored by formulation development scientists and engineers. It is a procedure by which product temperature at the sublimation interface may be measured during primary drying without placing any product in the vial. During the freeze-drying process, the valve between chamber and condenser is quickly closed, thereby isolating the freeze-drying chamber from the condenser for a short time. The MTM method records the pressure versus time data and analyzes the data to calculate the temperature at the sublimation interface.

By monitoring MTM as a critical process parameter, Pikal et al. were able to demonstrate that the exact product temperature heterogeneity is specific to the freeze-drying conditions, which can be minimized by applying thermal shields (i.e., empty vials around the sample vials and aluminum foil attached to the inside of the chamber door). In a system heterogeneous in product temperature, MTM measures a temperature close to the coldest temperature in the system. Finally, MTM provides a valid measurement of product temperature during primary drying even at temperatures as low as $-45°C$.

Mass balance determination using a tunable diode laser absorption spectroscopy (TDLAS) has recently been introduced as an analytical tool capable of obtaining real-time water vapor concentration and gas flow velocity measurements in the duct connecting the dryer chamber and condenser without process interruption [49, 51]. The device is used to determine the duration of the drying cycle to determine the total amount of water removed during the drying process. Pikal et al. [48] demonstrated that gravimetric and TDLAS flow sensor mass flux determinations showed a deviation of 5%, on average, during lyophilization cycles. With this flow senor, detection of primary and secondary drying end points was possible in real time, thus making the system amenable to end point analysis as a critical process parameter, rather than a time-based parameter.

12.5 PAT TOOLS FOR BIOPROCESS STARTING MATERIALS, DEFINED MEDIA, AND COMPLEX RAW MATERIALS

One of the most difficult, but the most fruitful, areas of controlling bioprocess variability pertains to incoming raw materials analysis for components used in bioprocesses. Too frequently, bioprocess manufacturers rely on vendor Certificates of Analysis to ensure the quality of the raw material they are using to manufacture their product. While many biopharmaceutical companies have QC raw material testing laboratories, the off-line analyses have to deal with many of the same issues as off-line product quality analysis. The analyses occur at a stage of the material manufacturing where it is too late to react to raw material quality changes that occur between the time of testing and the time of use. Often, the methods used are compendial, and while they faithfully detect the quality of the desired raw material, they are often not adequate enough to detect impurities and adulterants. Consequently, the finished product may result in rework, scrap, or even product safety issues.

Raw material stability may play a role in process variability for materials that are heat, moisture, or photosensitive, which may not be properly controlled. Analysis, just prior to use, is often a desired control point for most process scientists and engineers. Several spectroscopic techniques are suitable for in-line, "prior to use" raw material monitoring such as optical density, NIR, and fluorescence, which may allow process decisions to be made prior to the initiation of the manufacturing process. In addition, a host of off-line methods are being used to better understand how individual components of complex raw materials affect process outcomes.

One of the most abundant raw materials used in bioprocess unit operations is water. One in-line PAT tool to help with water monitoring is for total organic content (TOC). Online TOC is becoming a more commonplace tool for determining effects of carbon source from water on the process. Several commercial systems already available have the potential to reduce the sample load of QC labs to test for TOC.

For defined media, the components going into the defined media are measured and/ or controlled independently prior to mixing, but little attention is paid to interactions between components upon mixing. An increasing number of pharmaceutical and biopharmaceutical companies are using in-line or off-line spectroscopic instruments to perform raw material testing in their supply chain. The library approach using NIR is the most common in industry. The identity of a class of raw materials is usually confirmed with a spectral library of know "good" raw materials that have been confirmed through a use test.

As outlined by Reicht [46], the selection of samples is critical to the success of the application. Two sets of samples are required: one for the construction of the library and an independent one for external validation purposes to verify performance. The number of batches required to train the system depends on the required specificity of the method. The training set must collectively describe the typical variation of the substance being analyzed. Identification involves a smaller number of different batches (3–5) while qualification requires at least 20. With an appropriate calibration setup, NIR provides simultaneous quantitative measurements, such as moisture content and particle size determinations of raw materials. In-line probes such as OD, NIR, and fluorescence are placed into raw material or fed-batch media storage tanks where an assessment is made of the material prior to opening a valve to introduce the component or complex media into "fed-batch" fermentation or cell culture processes.

Other off-line techniques are also being used to gain a deeper process understanding of how individual components in simple and complex raw materials impact process variability. Techniques such as NMR, HPLC (with DAD or MS), and induction coupled plasma mass spectrometry (ICP-MS) are commonly employed for these purposes. The data are commonly correlated with a raw material "use test" to assess its impact on cell culture or fermentation critical process parameters (such as cell viability or product titer). These data are then coupled with chemometric visualization tools to more readily identify potential root cause of the raw material discrepancies. These types of in-depth insights into complex raw materials are beginning to create a paradigm shift on what level of analysis is needed to appropriately control process variability and product safety and know what additional testing must be performed prior to introduction into the manufacturing process. Many of the off-line analyses are amenable for conversion in-line analysis, if the process requires this level of control.

An example of an analysis of a complex raw material component is monitoring a vitamin breakdown product in minimum essential media (MEM) by UV analysis. Lanan and Kiistala [53] showed correlation of photostability of vitamins in a minimum essential media with cell culture productivity using two UV-based approaches. The first was an HPLC-based and the second used a 96-well plate-based approach. They demonstrated quantification of five vitamins in the MEM solution in addition to a vitamin photo-stability breakdown product. They were able to show that an increase in a breakdown product of one of the vitamins in the solution correlated with a loss of cell viability. Further research revealed that the breakdown product was, in fact, toxic to cells. With this proof-of-concept work, they determined that an optical density probe might be used to monitor vitamin solutions in the tanks prior to mixing with the other fed-batch media components. Such a system would allow engineers to test the material, just prior to use.

Scientists and engineers are also learning the importance of key trace elements contained in their more complex raw materials. Kiistala, Lanan, Houde, and Donegan [54] performed flow injection electrospray mass spectrometry on feed media and cell culture fluid and correlated an increase in a "marker" mass with the lack of a trace element needed by the cells. The mass correlated with a buildup of lactate, which is a known metabolic dead-end in the Krebs cycle. By determining the identification of the marker mass, they were able to isolate a portion of the Krebs cycle that relied on the presence of a certain trace element to keep the metabolic process from shunting to an anaerobic pathway. By adding more of the trace element copper (Cu^{++}) to the process, the overall improvement cell viability was realized.

Other tools used to assess complex raw materials can range from the simplistic fingerprinting approach to a complete breakdown of each component by multiple techniques. Hydrolysates from various animal and plant sources are regularly used in cell culture processes. A simple approach to assess hydrolysates is running reverse-phase HPLC with a diode array detector on a C-18 column and creating hydrolysate fingerprints from the spectra. The spectra from source material deemed "good" from use testing can be compared with incoming hydrolysates using pattern-matching software to elucidate any differences. Layers of information may be supplemented by using detectors in series, such as mass spectrometry, to identify the exact nature of the difference.

Still other raw materials require a complete breakdown to understand how multiple components can interact to affect cell growth. Wei and Lanan (data unpublished, [54]) used induction coupled plasma mass spectrometry, NMR, diode array HPLC spectra, and mass spectrometry coupled with chemometrics such as principal components analysis and PLS-based visualization tools to demonstrate how components in hydrolysates such as trace elements, hormone analogues, and key peptide quantities impacted cell viability. They were able to determine by ICP-MS the varying levels of the trace element zinc in different lots of hydrolysate, but this component alone was not responsible for all of the variability. They used NMR and diode array HPLC spectra to further determine the presence of certain hormone analogues and peptides in each lot of hydrolysate. By putting the data together and ranking combinations of each component based on effect on cell viability, they were able to put together a matrix of component interactions that described how each component and combinations of components were correlated with cell growth.

With the ability of raw materials to have the largest impact on process variability, it is becoming increasingly important to gain a deeper fundamental understanding of how each component influences critical process parameters and product quality attributes including post-translational modifications such as glycosylation. As raw materials become more defined and characterized, the ability to selectively influence product quality attributes during the course of a bioreactor run becomes possible. NIR can also be used for resin identity as a quality check before column-packing procedures are initiated.

12.6 CHEMOMETRICS AND ADVANCED PROCESS CONTROL TOOLS

Aforementioned sampling techniques, new sensor technologies, and new analyzers generate complex data that require special treatment to extract the embedded useful information content. Chemometrics is the science of extracting the relevant information that is coming from various chemical sensors and analyzers by applying advanced mathematical and statistical algorithms. It is commonly employed especially when spectral measurement systems are used in a given process. For instance, in a bioreactor setting where off-line or online samples are taken for conducting an NIR and/or HPLC (or an off-gas GC analysis) in certain cultivation periods (e.g., every 12 or 24 h), resulting data set is fairly complex such that it requires pretreatment (such as normalizing, scaling, denoising, and outlier removal) and its relevant information to be extracted at each time point of sampling (e.g., NIR wave numbers that correlate with certain metabolites). This type of spectral data sets may be noisy—there may be missing data points and outliers— and highly collinear, which makes conventional analytical techniques (such as multiple ordinary linear regression) inappropriate to explain these data (conventional techniques become mathematically instable when presented with this type of data). Chemometrics tools include, but are not limited to, principal components analysis (PCA), partial least squares (PLS), continuum regression, evolving factor analysis, and principal components regression (PCR). PCA and PLS are by far the most used techniques for practical applications and are now available via commercial off-the-shelf (COTS) software. Another major advantage of using chemometrics techniques such as PCA and PLS is that while explaining the complex data sets, these techniques mathematically reduce the dimensionality of the system so that only the relevant information is extracted by using a few variables instead of many hundreds or thousands. PCA does this by finding the major variation directions and PLS by also correlating these directions with certain output (response) variables that are thought to be impacted by many inputs. A short summary of both techniques is provided in the following sections.

12.7 THE POWER OF PLS AND PCA

A large number of variables, interactions among the variables, and complex process dynamics (such as biological dynamic behavior) pose challenges against analyzing the process data involved in the analysis. Multivariate data modeling technology is used to reduce dimensionality of the problem as well as handling colinearity and missing data. When only the input space or output space is to be analyzed or just the major variability differences in a given process or system is of concern (not necessarily the input–output relations), PCA is employed. PCA decomposes the covariance matrix of the data set (in a linear fashion) to explain the maximum variability and remove noise in the process. Therefore, it finds the major variability directions within the data set to explain the overall system behavior with a few components as shown in Fig. 12.7. In this example, measurements are made on three variables for many times (e.g., across batches, repeats, or multiple spectra) depicted by the points that form a data volume or cloud stretched toward some directions based on the process or system characteristics and measurement systems variability. PCA finds the first major variability direction (also called principal component one, PC_1), removes this explained variation from the data set (working on the residuals after the first iteration), and looks for the second major variability direction (PC_2). As shown in Fig. 12.7, a three-variable process can be explained by one or two

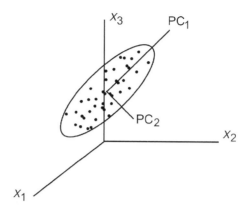

Figure 12.7. PCA representation of a three-variable process.

PCs. When there are many more variables, this becomes a great advantage as it enables summarizing overall variability with a few variables. Once a PCA model is developed, it can be used to quickly identify separations in the data set, if there is any, and one can relate this difference to original variables that are used in the model.

One example is the raw material data (related to raw material constituents, such as physical and chemical characteristics); when the data are available for different lots of the same raw material, one can quickly analyze if there is any major difference between them with respect to the variables measured (usually there are many variables measured including the spectral data) as shown in Fig. 12.8. The 3D-score plot on the left indicates a summary of the entire raw material variability (defined by the confidence volume of the ellipsoid) for all of the measured variables (in this example, both physical, such as particle size and filterability, and chemical characteristics, such as amino acids, are used). Measurements from a new raw material lot are projected onto this PCA model for quick comparison that shows no major difference overall. However, some variables such as the number of amino acids are found slightly different by studying the contribution plot on the right-hand side of Fig. 12.8.

When there is a need to explore multivariate correlations (say, input variables or predictors) against one or more response variables (i.e., outputs or predictees), a variety of regression techniques are used in chemometrics.

Colinearity exists when two variables change in the same direction. ordinary linear regression (OLR) techniques fail to handle this type of complex data due to ill-conditioned covariance matrices causing an inability to take the inverse of the solution matrix leading to regression coefficients. Various alternative regression techniques have been developed and offered by chemometrics research community in the past 20 years that can be referred to as biased regression, reduced-rank regression, and subspace modeling regression. Some of the techniques that are widely used include principal components regression, partial least squares, and ridge regression. We will cover one of them that is widely used in practice and called partial least squares (or projection to latent structures).

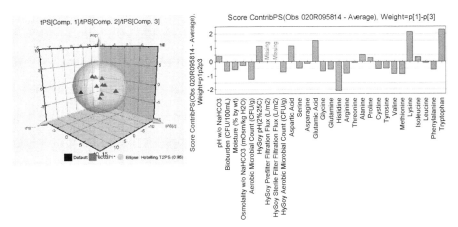

Figure 12.8. PC score 3D plot (top) showing various grades of raw materials and variable contribution plot (bottom) indicating the raw material constituent differences between the historical lots (black triangles) and a new lot (red triangle). (See the insert for color representation of this figure.)

PLS is a linear empirical modeling technique that reduces model dimensionality while retaining the useful information in the data. PLS rearranges process variables space (denoted as X) and response variables space (denoted as Y) by reducing their size (in terms of number of actual variables) while maximizing the covariance between them. For instance, when the three process variables (X_1, X_2, and X_3) depicted in Fig. 12.7 and one response variable (y) are to be correlated, the PLS model reduces the original predictor variable dimensions into a selected few latent variables (LAs) that are basically defined as weighted linear combinations of process variables (predictors). First latent variable (LV_1) is selected as the direction of maximum variability of X space that is most predictive of the y space. Once the first LV is determined, amount of variance it can explain is removed from the X and y spaces, and a second iteration is carried out by using the residuals of X and y to determine the second LA direction. Therefore, the second (LV_2) latent variable is selected as the direction of the remaining maximum variability of X space that is most predictive of the residual y space. In this example, most of the variation in the data set can be explained by using two LVs instead of three process variables.

The model dimensionality or the number of latent variables used in each model is determined by cross-validation. In cross-validation, each batch is removed from the data set once and a new PLS model is developed by using the remaining batches. The response variables (Y) of the left-out batch are then predicted by the model and compared with the actual values. A number of cross-validation statistics (including predicted residuals sum of squares (PRESS)) can be then calculated to determine the model dimensions (i.e., number of latent variables) and model's power in explaining variance [34, 35].

The following multivariate statistics and charts are typically used during exploratory analysis, monitoring, and prediction [58, 59] of batch/fed-batch processes.

Scores (t_1, t_2, ...) are the weighted summary of all the process variables used in the model. The score biplots represent the operational space and each point in the plot represents a single batch. The batches that are close to each other in the score biplots can be concluded as operated in a similar operating space. These plots enable determining if any group, pattern, or outlier exists in the data set.

Score contribution plots show why an observation in a score plot deviates from the reference point. The reference point can be the average behavior of the process, another observation, or a group of observations. Score contributions are typically depicted as bar plots. The dominating bars indicate the variables that show maximum deviation from the reference point. The direction of the bars indicates in which direction the variables deviate. The cause of the deviation can be determined by using the contribution plots and the process knowledge.

Loadings plots (w^*c) show both the X weights (w^*) and Y weights (c) superimposed in one plot. The (w^*c) plot of one PLS component against another (e.g., 1 versus 2) shows how the X variables correlate with Y variables and the correlation structure of the X's and Y's. To interpret this plot, one can imagine a line passing through the origin and one of the Y variables and project all the X and Y variables onto this line. The X variables near the Y variables are positively correlated with Y and the X variables that are on the opposite side of the Y variables are negatively correlated with Y. Also, the Y variables close to each other are positively correlated and the ones on the opposite sides are negatively correlated with each other.

Variable of importance in projection (VIP) plots summarize the importance of the variables both to explain X and to correlate with Y. VIP values larger than 1 indicate "important" X variables predicting Y.

The PLS regression coefficients show how strongly the Y variables are correlating with X variables. The bars indicate the confidence intervals of the coefficients. The directions of the bars show if the X variables are positively or negatively correlated with the Y variables. The coefficients are considered significant if the confidence interval does not cross zero. One should note that these coefficients are usually not independent unless generated by design of experiments.

12.7.1   Use of PCA and PLS in Multivariate Statistical Process Monitoring

Construction of multivariate charts (as explained above) offers multivariate statistical process monitoring (MSPM) capabilities for efficiently monitoring batch biopharmaceutical (they are also used in other batch industries) processes. Typical approach includes selecting a representative historical data set from a production database. It can be at various scales (i.e., bench, pilot, or large). However, availability of representative historical data is more probable at large scale due to routine production campaigns comprised of many batches. Usually, 15–30 batches are good numbers for constructing a reasonable process models via PCA and/or PLS. It is recommended that batch selection for modeling is carried out via rational subgrouping if the main objective is monitoring of successive batches. On the bass of our experience, rational subgrouping should be combined with the process and characterization knowledge. Once the historical representative batch data set is selected, outliers removed, missing data are treated, and data pretreatment (such as autoscaling) is done, one may deploy PCA and PLS algorithms (when applied to batch processes, these are also called multiway PCA and PLS [60–63]) to develop an empirical process model that is to represent nominal behavior of the process variability and its expected impact on output variables. This nominal model (or reference model or baseline model) is then used in monitoring new batches by comparison. When aforementioned MSPM charts are used (off-line or in real time) to monitor new batch performance against this historical norm (or baseline), deviations can be detected and responsible variables can be diagnosed within this advanced monitoring framework. A thorough analysis and methodology on how to develop an MSPM framework particularly in biopharmaceutical industries can be found in the literature [64–67].

As an industrial example, two consecutive bioreactors are modeled via MPLS to devise a real-time MSPM (RT-MSPM) framework. In this setting, historical batches are mined from the manufacturing databases to develop nominal process models for each bioreactor. Online quality prediction models (mathematical models predicting the end process quality such as final titer) are also developed for providing early estimation of quality attributes [68].

A comprehensive multivariate process monitoring and final quality estimation are made possible in this real-time framework above. As shown in Fig. 12.9a, the time series traces of the first score for two bioreactors (first two charts cover the batch and fed-batch phases of a seed bioreactor, respectively, followed by the score time series chart for a production bioreactor). These charts are useful in detecting deviations from nominal operation by taking many variables into account together at each time instance (based on the data sampling frequency from the measurement systems on the bioreactors). The red lines are 95% confidence limits indicating the multivariate in-control region; the green curve is the mean (or average) curve, denoting the nominal behavior when the batch is run at average trajectories. The dark colored curve of the chart in the middle (of Fig. 12.9a, i.e., the fed-batch phase of the seed bioreactor) is for monitoring a new batch while it is in progress and shows signs of deviation toward the halfway of the run. The same deviation is also detected by the Hotelling's T^2 chart in Fig. 12.9b, process maturity progress chart in Fig. 12.9c, and Squared Prediction Error (DMoDX) chart in Fig. 12.9d. Once the deviation is detected, one can identify the variable or variables that may have caused this deviation by using the contribution plots. Interactively clicking the scores, T^2, or DMoDX chart where the deviation point is, respective contribution plot can be generated as shown in Fig. 12.10. Finally, we can go into the variable level from the contribution plot interactively. In this example, variables 1 and 2 seemed to cause the multivariate statistics to inflate and signal a deviation in the MV charts in (Fig. 12.9a–d). Actual time series plots for these variables indicate a clear deviation from their average trajectories. Root cause investigation revealed that this deviation was due to a temporary power failure. As shown in this example, by just monitoring a few MV charts (Fig. 12.9), engineers and scientists can avoid having to monitor many different charts (as many as 20–30 in this example) and quickly identify the deviations within the MV framework (Fig. 12.10).

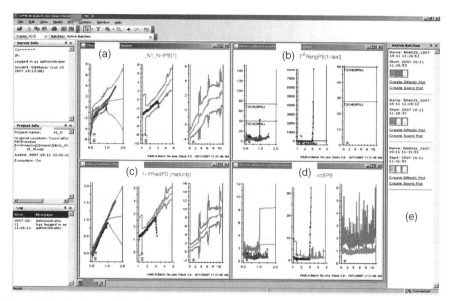

Figure 12.9. Online real-time detection and diagnosis of a process deviation enables engineers to troubleshoot the process more effectively. Panel (a) score time series plots, (b) T^2 time series plots, (c) batch maturity plots, (d) DModX (or squared prediction error) plots, and (e) executive batch dashboard for two consecutive bioreactors (first two plots or boxes in the dashboards for two phases in the first bioreactor, batch and fed-batch). (See the insert for color representation of this figure.)

Note that the dashboard shown in Fig. 12.9e also indicates (color codes) the fed-batch phase of this particular batch as out-of-multivariable control. Therefore in essence, one can simply monitor the dashboard to get the executive summary of process in-control state in real time and focus on the deviation by interactively exploring the area of focus that is needed due to any deviation. This hierarchical approach not only simplifies engineers and operators' supervision tasks but also makes them more efficient in a proactive manner.

Besides using multivariate approaches, a number of soft-computing techniques have also been developed to monitor biopharmaceutical manufacturing process performance. Examples include using the existing measurements to derive calculated variables or statistics to achieve increased sensitivity in monitoring performance and obtain predictions. In addition, practical approaches are proposed to overcome data alignment problems when measurements are made only with respect to time. One common approach includes monitoring the progress of a batch with respect to an indicator variable that can also be considered as relevant time variable so that the rest of the process variables can be monitored against its progression points over time. This process variable is selected to indicate the progress of a batch instead of time. Each new measurement is then sampled relative to the progress of this variable. This variable, which indicates the

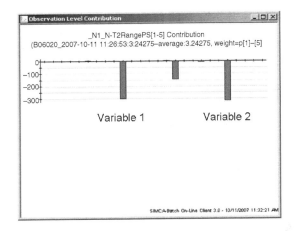

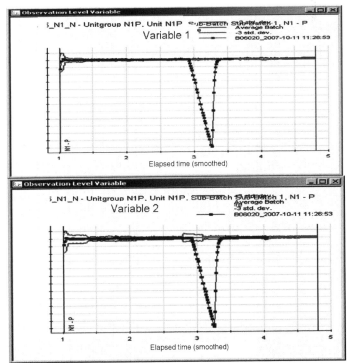

Figure 12.10. Contribution plot (top) and actual variable time series charts (bottom) for variables 1 and 2 against their mean trajectories and ±3 SD variability ranges.

"relevant time," should be smooth, continuous, monotonically increasing (or decreasing), nonnoisy, and spanning the range of all other process variables measured within the same unit procedure of interest [65]. A practical application of this approach in biopharmaceutical purification processes is proposed by Larson et al. [37] for monitoring the chromatography column packing integrity as explained in more detail in the following section.

12.8 "RELEVANT TIME" COLUMN INTEGRITY MONITORING (MOMENTS ANALYSIS VERSUS HETP)

Monitoring the quality of packing and detecting integrity breaches in purification liquid chromatography columns is important from process performance and final product quality perspective. Conventional techniques for this task include monitoring a few key variables such as step yields, prepool volumes, and pulsed-injection height equivalent of a theoretical plate (HETP) testing [68, 69]; or simply inspecting the chromatograms visually. Transition analysis uses existing process data (solute signals such as pH, conductivity, and UV) during buffer transitions at the column outlet during a chromatography operation to assess the packed bed integrity. The postcolumn signal (e.g., conductivity) is noise filtered and normalized, and its derivative is taken to calculate several parameters that are indicative of the packed bed integrity. These calculated parameters are orthogonal and include the following:

1. Number of inflection points, N
2. Maximum rate of change or maximum derivative $(dC/dV)_{max}$
3. V_{max}
4. Breakthrough volume (BTV)
5. Cumulative error or peclect number
6. Non-Gaussian HETP
7. Gaussian HETP
8. Asymmetry factor
9. Skewness
10. Kurtosis
11. Overall integrity

These transitional measures are much more sensitive than pulse injection-based HETP and asymmetry results enabling an early detection of column integrity deterioration during its use at each cycle. Depending on the chromatography column type, there are multiple opportunities to perform transition analysis within the process phases including equilibration, elution, and the clean to store step.

12.8.1 Number of Inflection Points

The number of inflection points during a transition is determined by counting the number of peaks from a plot of (dC/dV) versus V. All those peaks higher than 10% of $(dC/dV)_{max}$ are counted. An increase in the number of inflection points indicates early breakthrough that is associated with an integrity breach.

12.8.2 Maximum Rate of Change

The maximum rate of change is the maximum value of $(dC/dV)_{max}$. Decreasing $(dC/dV)_{max}$ over multiple uses indicates transition spreading over a larger volume that is indicative of integrity breaches.

V_{max} is the corresponding volume to the $(dC/dV)_{max}$. Any change in the location of the maxima during transition is indicative of changes in bed integrity.

12.8.3 Breakthrough Volume

The breakthrough volume corresponds to the first value of V at which the normalized solute signal is greater than 0.05 (for a step-up change). A decrease in the breakthrough volume, that is, early breakthrough is a characteristic of integrity breaches.

12.8.4 Cumulative Error

The cumulative error is the difference between the actual solute signal and the signal expected from a noninteracting plug flow. Factors that increase dispersion such as column headspace or integrity breaches result in higher cumulative error.

$I_i = \int ABS(C\text{-}0)dV$ between $V(0)$ and $V(0.5)$
$I_f = \int ABS(1\text{-}C)dV$ between $V(0.5)$ and $V(1.0)$
Cumulative error $E = I_i + I_f$

12.8.5 Non-Gaussian HETP (HETP$_N$)

The non-Gaussian HETP is calculated by the following formula:

$$HETP_N = L\sigma^2/(M_1/M_0)^2$$

where L is the length of the column, σ^2 is the variance, and M_1 and M_0 are first and zeroth moments, respectively. The moments equation for kth order is given by $M_k = \int V^k(dC/dV)$ dV. The variance for the distribution is calculated from the zeroth, first, and second moments by using the following equation:

$$\sigma^2 = (M_2/M_0) - (M_1/M_0)^2$$

This estimate of theoretical plates takes into account the entire transition without assuming a Gaussian distribution. An increase in HETP$_N$ value correlates with decreased efficiency of the column.

12.8.6 Gaussian HETP (HETP$_g$)

HETP$_g$ expresses column efficiency as a ratio of the bed height to the theoretical number of plates. The calculation estimates plate height with the assumption that the solute response peak is a Gaussian distribution. The following formula is used to calculate

HETP$_g$:

$$\text{HETP}_g = \frac{\text{Bed height}}{5.54^* \left(\frac{V_{max}}{V_d - V_c}\right)^2}$$

where $V_{max} =$ the number of column volumes from the transition initiation to the maximum rate of change, $V_d =$ CV at 1/2 peak max (right), and $V_c =$ CV at 1/2 peak max (left).

12.8.7 Asymmetry

Asymmetry factor is a measure of skewness at 10% of peak maximum. The equation is as follows:

$$A_f = \frac{V_b - V_{max}}{V_{max} - V_a}$$

where $V_{max} =$ the number of column volumes from the transition initiation to the maximum rate of change, $V_a =$ CV at 10% peak max (left), and $V_b =$ CV at 10% peak max (right).

12.8.8 Skewness and Kurtosis

Recently, skewness and kurtosis are also added to transition analysis measures as higher moments with some level of success. They provide higher sensitivity; however, they should be used with care and appropriate signal conditioning to avoid false alarms. Conley and Thommes [70] have reported positive results for using skewness not only to discern change types in column integrity but also to demonstrate removal of impurities.

- *Skewness:* A measure of the *asymmetry* of the distribution
- *Kurtosis:* A measure of the *flatness* of a distribution

C_i, normalized solute signal (e.g., conductivity, pH) value at ith volume reading; V_i, normalized buffer volume at ith volume reading; V_R, normalized buffer volume at the inflection point of V_i versus C_i; σ, sample standard deviation of solute signal; and n, number measurements based on buffer volume.

For a step increase transition (as shown in Fig. 12.11),

$$C_{max} = \sum_{i=1}^{n} \Delta C_i = 1, \qquad V_R \approx \sum_{i=1}^{n} V_i \cdot \Delta C_i, \qquad \sigma^2 = \sum_{i=1}^{n} (V_i - V_R)^2 \Delta C_i$$

$$\text{Skewness} = \left[\frac{\sum_{i=1}^{n} (V_i - V_R)^3 \cdot \Delta C_i}{\sigma^3}\right], \qquad \text{Kurtosis} = \left[\frac{\sum_{i=1}^{n} (V_i - V_R)^4 \cdot \Delta C_i}{\sigma^4}\right]$$

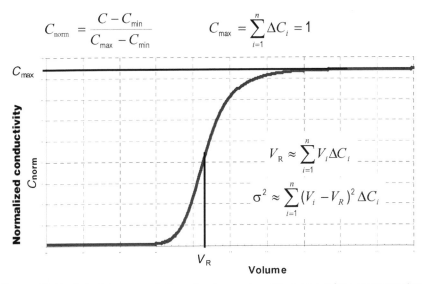

$$C_{norm} = \frac{C - C_{min}}{C_{max} - C_{min}} \qquad C_{max} = \sum_{i=1}^{n} \Delta C_i = 1$$

$$V_R \approx \sum_{i=1}^{n} V_i \Delta C_i$$

$$\sigma^2 \approx \sum_{i=1}^{n} (V_i - V_R)^2 \Delta C_i$$

Figure 12.11. Pictorial representation of a step-up transition on a solute concentration.

12.8.9 Overall Integrity Measure

The overall integrity measure is a weighted compilation of the results from the 10 individual transition analysis parameters. It allows a single parameter to be trended. To normalize the dimensions of each parameter, the percentage of the average is used to measure the offset of each parameter from its historical average. Each percentage of average is weighted based on its individual sensitivity to detecting integrity failures. The values are then summed and averaged together to yield the overall integrity measure:

$$\% \text{ of average} = \frac{\text{Parameter value}}{\text{Historical average}} \times 100$$

$$\text{Overall integrity} = \frac{\sum [\text{weight} \times (\% \text{ of average} - 100\%)]}{\sum \text{weight}}$$

Industrial examples as shown in Fig. 12.12 indicate the efficiency of transition analysis against conventional techniques in detecting salient changes in column integrity and performance. Lee et al. [36] have shown that the transition response curve after 12 column uses indicates a prepeak, a potential issue, early warning to operations, whereas this is not easily discernable by inspecting UV elution profile. Similarly, monitoring transition analysis measures over time (or use between packing cycles as

UV elution profiles do not show column integrity changes

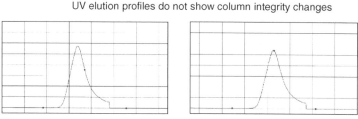

Elution UV profile after 3 column uses   Elution UV profile after 12 column uses

Transition analysis reveals column integrity changes

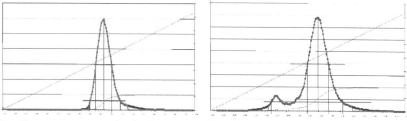

Transition response curve after 3 column uses   Transition response curve after 12 column uses

Figure 12.12. Monitoring UV elution profiles versus transition analysis results [36].

shown in Fig. 12.13) provides an effective and sensitive means of detection of adverse trends in column integrity changes that may lead to process upsets and product quality issues [36].

12.9 CHALLENGES FOR IMPLEMENTATION OF PAT TOOLS

12.9.1 Organizational Structure

The challenges of implementing PAT in bioprocess manufacturing is always sure to elicit vigorous discussion and debate within any biotechnology company involved in a PAT initiative. In terms of governance, typically, a hierarchical matrix structure exists within biologic manufacturing organizations. The matrix consists of individual PAT project teams, a project governance committee, and an executive governance board. The teams are generally made up of a group of employees operating cross-functionally. They are responsible for identifying the process problem (or improvement opportunity), brainstorming on possible solutions, designing and performing a proof-of-concept study to test out the potential PAT tool, risk assessment, and finally bringing it to the governance committee for discussion, review, and decision making as to whether the project should be advanced (go/no-go). The governance committee typically consists of senior management personnel from process development, manufacturing, and quality that assess a project's merit and determine if the technology is better suited for implementation in the

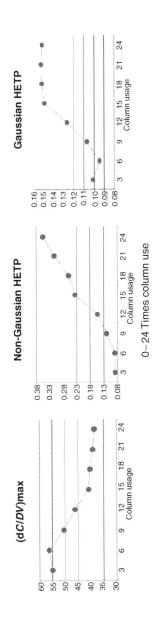

Figure 12.13. Monitoring the packing performance with a few transition analysis measures over number of packings [36].

development environment only or if it has potential to be implemented in a manufacturing environment. The executive governance board generally procures resources, both capital and manpower, to ensure the success of the projects.

12.9.2 Organizational Management Challenges

There are many pitfalls to consider when setting up a PAT initiative within a biopharmaceutical company. One scenario to be aware of is related to the interaction between the PAT teams and the governance committee. Typically, PAT teams are interested in fixing "real" problems on the manufacturing floor that resolve pain points related to high process variability. PAT teams put large amounts of time and energy into developing creative solutions and tools. When the tool is proposed to the governance committee and determined to be nonvalue added by the committee, it can leave the team reluctant to propose future solutions. Some of the answers can be found in the implementation of tools found in Six Sigma teams that empower the team members to make the decision to terminate a project or continue the project with agreed upon deliverables back to the "sponsors" (governance committee).

Silo thinking between departmental lines is another pitfall to be aware of when implementing PAT initiatives. Process developers tend to consider themselves as the process owners, as they typically have the most experience with the process when it is first introduced into the manufacturing environment. As a result, once the process has been transferred into manufacturing, with enough runs, manufacturing personnel feel that they now own the process. When one department proposes a PAT tool to add to the process to fix a "problem," it can be viewed as a threat in the other department. The process developers may disagree that the process needs improvement because the development was rigorous and shown to be scaleable and easily validated. The manufacturing department may disagree that an additional tool is needed because they have the most experience that allows them to operate the process without additional controls. Early consensus building, from the conceptualization stage, is imperative to avoid falling into a silo mentality. Early involvement of personnel from each group impacted by the proposed change, allows the team to proceed forward and create the "buy-in" so that all parties agree on the merit of a proposed project.

Another key area for a company to focus attention upon when considering PAT implementations is to engage quality and regulatory departments early and often. Encouragingly enough, a result from a 2004 PPAR (Pharmaceutical Process Analytics Roundtable) survey noted that regulatory and QA "buy-in" were becoming less of a barrier for introducing PAT technologies. In fact, it was noted that organizational commitment to PAT as a process development effort was the main barrier cited in the survey.

Regardless, regulatory and quality strategies must be developed to introduce the tools into the manufacturing process and collect information, at scale, to determine if the data provide the expected information needed to control the process. This is typically managed through protocols that define the scope and duration of the evaluation and what decision criteria will be involved to change the process to include the technology. As Rathore and others point out, regulatory strategies need to be developed to address concerns agencies may have when looking at parameters such as step yield, which have historically been viewed as critical process parameters. Well-defined data sets and risk analysis should be able to justify replacing "historical" process parameters with ones based on end point analysis or product quality attribute decisions. These types of decisions will need to be addressed in the appropriate sections of regulatory filings as well as other company documentation such as batch records and SOPs.

12.10 FUTURE PAT TOOLS

PAT tools are taking longer than some practitioners would like, but it is becoming more ingrained in corporate philosophies and operational excellence thinking. Many practitioners struggle with the lack of availability of the appropriate analytical tools to get the information needed to obtain the desired "window" into the process. Many companies struggle with issues of intellectual property and funding collaborations with instrument manufacturers who are willing to custom fabricate tools to fit their manufacturing processes. In spite of these types of barriers, the need to learn more about biological manufacturing processes is still driving innovations in academic, research institutes and industry geared to arriving at new and improved tools for monitoring and controlling the processes.

Enabling tools and technologies that may provide new ways of looking at biological processes in the future may include techniques such as ultrasonic detection, nanofluidics and SPR arrays, automated microimaging techniques, automated flow injection analysis, calorimetry, and rugged spectroscopic techniques to address loss on filtration and loss on drying for monitoring critical quality attributes in cell culture fluid. Further refinement and ruggedness testing is needed in the instrument development arena end to address ease of use and ease of implementation in a manufacturing environment.

Ultrasonic detection of protein aggregates in cell culture broth may allow improved strategies and controls to be developed for harvest steps and will certainly help properly load columns for initial purification steps. Since the technology does not rely on light scatter to quantify process changes, it allows implementation of the tool to evaluate opaque solutions. In fact, ultrasonic reflectometry is already being used to assess membrane integrity in microfiltration and ultrafiltration skids [55]. This technology could also be used to assess slurry uniformity prior to column pouring and packing. As the ultrasound technology becomes better understood and data deconvolution algorithms improve, this may become a key window into our processes that does not exist today. Sound waves are also being used to generate equipment performance process signatures for applications such as granulation [71, 72] and fluidized bed coating processes [73] that allow engineers to proactively perform preventive maintenance before an event occurs that could potentially result in a lost batch upon failure. Clearly, sound wave technologies are destined to be one of the tools in the PAT toolbox of the future.

Online monitoring bioreactor monitoring by calorimetry [56] may be another PAT tool in the future that scientist and engineers will begin to rely upon to make process decisions. The simple fact that heat production in combination with the material flows correlated with metabolic maps may allow finer control of cellular metabolic processes is an incentive to develop this technology further. It has the potential to provide real-time information about growth kinetics and stoichiometry when harnessed with appropriate chemometric tools.

Another area ripe for further development is further automation of flow injection analysis for biopharmaceutical purification skids as a means for real-time feedback loops. Since purification step runs are generally measured in hours, a simple and rapid technique for quality attribute assessment in a 2 min time frame may allow implementation of these technologies in future purification processes.

In the arena of formulations, technologies that enable relevant time analysis of compounding steps, loss on terminal filtration and loss on drying, should be investigated. For facilities and plant monitoring, built-in analytical redundancy and tools such as acoustic spectroscopy to measure the health of unit operation components in a plant should be further explored. Data infrastructure improvements need to be made that should include further cross-industry harmonization on unified connectivity language, such as AniML (analytical mark up language) and OPC (open connectivity) for plant and analytical instrumentation to allow data modeling and process control implementations.

While some practitioners of PAT may yearn for more immediacy, the process of adopting PAT will likely follow the timeline of getting a new drug approved. As most scientists, engineers, and managers in the biotechnology have experienced, the drug development process leading to filing of a Biologics License Application, this can take 5–10 years. Many biopharmaceutical companies are reluctant to implement these tools in existing commercial processes for fear of putting their revenue stream at risk. Rather, implementation of these tools will more likely be a gradual process and be put in place as new processes are developed.

ACKNOWLEDGMENTS

Michael Molony would like to acknowledge his former colleagues at Biogen Idec, Marissa Braganza, Eddie Rustandi, Yao-Ming Huang, Eric Rosemyer, Maureen Lanan, Mia Kiistala, Kip Lowery, Alime Kirdar, Damian Houde, Justin McCue, Shefal Parikh, Mylene Talabardon, Kelly Wiltberger, Robin Hyde-Deruysher, Joydeep Ganguly, Michael Villacorte, Mahalia Ong, Soheil Rhamati, Valerie Tsang, Rohin Mhatre, Jorg Thommes, David Chang, and Thomas Ryll, as well as a current colleague from Allergan, Ron Bates, who graciously provided editorial services; and finally, his collaborators Jun Park, Kurt Brorson, Bob Mattes, Rick Cooley, George Barringer, and Mel Koch.

Cenk Undey would like to acknowledge his former colleagues at Amgen, Hiren Ardeshna, Roderick Geldart, Asti Goyal, Wenshan Lee, Marty Martin, Julie Matthews, Jay Stout; at FDA, Ali Afnan; his current collaborators and management at Amgen, Sinem Ertunc, Manuj Pathak, Katherine Chaloupka, Erin Collins, Sourav Kundu, Thomas Mistretta, James Stout, Douglas Inloes, Dane Zabriskie, Kimball Hall, Tony Pankau, Chris Bush, Marion Baust-Timpson, Sean Kelly, Alice Gardiner, Letha Chemmalil, Steven Hunt, Robert McInerney, Larry Cater, Scott Kendra, Jayne Morris, Alex Lee, Chris Vales, Deniz Bac, Seth Moye, Murali Pasumarthy, Duncan Low, and Joseph Phillips. Finally, thanks to Fetanet Ceylan Undey of Amgen for her support and editorial comments.

REFERENCES

[1] Morrow KJ Jr. Optimizing antibody biomanufacture: better processes can consistently yield both quality and quantity. *Genet Eng News* 2007;27(4):34–35.

[2] Ge Z, Cavinato AG, Callis JB. Noninvasive spectroscopy for monitoring cell density in a fermentation process. *Anal Chem* 1994;66:1354–1362.

[3] Vaccari G, Dosi E, Campi AL, Mantovani G, Gonzalez-Vara A, Matteuzzi D. A near-infrared spectroscopy technique for the control of fermentation processes: an application to lactic acid fermentation. *Biotechnol Bioeng* 1994;43:913–917.

[4] Yeung KSY, Hoare M, Thornhill NF, Williams T, Vaghjiani JD. Near-infrared spectroscopy for bioprocess monitoring. *Biotechnol Bioeng* 1999;63:684–693.

[5] Arnold SA, Matheson L, Harvey LM, McNeil B. Temporally segmented modeling: a route to improved bioprocess monitoring using near infrared spectroscopy? *Biotechnol Lett* 2002; 23:143–147.

[6] Macaloney G, Hall JW, Rollins MJ, Draper I, Anderson KB, Preston J, Thompson BG. The utility and performance of near-infrared spectroscopy in simultaneous monitoring of multiple components in a high cell density recombinant *E. coli. Bioprocess Eng* 1997; 17:157–167.

[7] Arnold SA, Gaensakoo R, Harvey LM McNeil B. Use of at-line and *in-situ* near-infrared spectroscopy to monitor biomass in an industrial fed-batch *Eschericha coli* process. *Biotechnol Bioeng* 2002;80:405–413.

[8] Braganza M, Rustandi E, Huang Y-M, Molony M. On-Line Amino Acid Analysis from a Small Scale CHO Cell Bioreactor. Agilent Application Bulletin No. 5989-7240EN (August), 2007.

[9] Baylor LC, O'Rourke PE. UV-Vis for on-line analysis. In: Bakeev KA, editor. Process Analytical Technology: Spectroscopic Tools and Implementation Strategies for the Chemical and Pharmaceutical Industries. Blackwell Publishing Ltd; 2005. p 171–186.

[10] Lee HLT, Boccazzi P, Gorret N, Ram RJ, Sinskey AJ. *In situ* bioprocess monitoring of *Escherichia coli* bioreactions using Raman spectroscopy. *Vib Spectrosc* 2004;35:131–137.

[11] Rathore AS, Yu M, Yeboah S, Sharma A. Case study and application of process analytical technology (PAT) towards bioprocessing: use of on-line high-performance liquid chromatography (HPLC) for making real-time pooling decisions for process chromatography. *Biotechnol Bioeng* 2008;100:306–316.

[12] Mattes RA, Root D, Chang D, Molony M, Ong M. *In situ* monitoring of CHO cell culture medium using near-infrared spectroscopy. *BioProcess Int* 2007;5 (January Suppl.1): 46–49.

[13] Arnold SA, Crowley J, Woods N, Harvey LM, McNeil B. *In-situ* near infrared spectroscopy to monitor key analytes in mammalian cell cultivation. *Biotechnol Bioeng* 2003;84:13–19.

[14] O'Haver TC, Begley T. Signal-to-noise ratio in higher order derivative spectrometry. *Anal Chem* 1981;53:1876–1878.

[15] Arnold SA, Harvey LM, McNeil B, Hall JW. Employing near-infrared spectroscopic methods of analysis for fermentation monitoring and control, Part 1. *Bio Pharm* 2002;26–34.

[16] Montague G. *Monitoring and Control of Fermenters*. Rugby, UK: Institute of Chemical Engineers; 1997.

[17] Card C, Hunsaker B, Smith T, Hirsch J. Near-infrared spectroscopy for rapid, simultaneous monitoring of multiple components in mammalian cell culture. *BioProcess Int* 2008;6(3): 58–65.

[18] McShane MJ, Cote GL. Near-infrared spectroscopy for determination of glucose, lactate, and ammonia in cell culture media. *Appl Spectrosc* 1998;52(8):1073–1078.

[19] Wu P, Ozturk SS, Blackie JD, Thrift JC, Figueroa C, Naveh D. Evaluation and applications of optical cell density probes in mammalian cell bioreactors. *Biotechnol Bioeng* 1995;45: 495–502.

[20] Schmid G, Zacher D. Evaluation of a novel capacitance probe for on-line monitoring of viable cell densities in batch and fed-batch animal cell culture processes. In: Animal Cell Technology Meets Genomics, Chapter 5. Netherlands: Springer; 2005. p 621–624.

[21] Cooley RE, Stevenson CE. On-line HPLC as a process monitor in biotechnology. *Process Control Qual* 1992;2:43–53.

[22] Ansorge S, Esteban G, Schmid G. On-line monitoring of infected Sf-9 insect cell cultures by scanning permittivity measurements and comparison with off-line biovolume measurements. *Cytotechnology* 2007;55:115–124.

[23] Hall JW, Rollins MJ, Macaloney G. Analyzing bioprocess systems using near-infrared spectroscopy methods. *Genet Eng News* 1994;14:18.

[24] Hall JW, McNeil B, Rollins M, Draper I, Thompson BG, Macaloney G. Determination of nutrient levels in a bioprocess. Near-infrared spectroscopic determination of acetate, ammonium, biomass and glycerol in an industrial *Escherichia coli* fermentation. *Appl Spectrosc* 1996;50:102–108.

[25] Brimmer PJ, Hall JW. Determination of nutrient levels in a bioprocess using near infrared spectroscopy. *Can J Spectrosc* 1993;38:155–162.

[26] Larson TM, Gawlitzed M, Evans H, Albers U, Cacia J. Chemometric evaluation of on-line high pressure liquid chromatography in mammalian cell culture: analysis of amino acids and glucose. *Biotechnol Bioeng* 2002;77:553–563.

[27] Jestel NL. Process Raman spectroscopy. In: Bakeev KA, editor. Process Analytical Technology: Spectroscopic Tools and Implementation Strategies for the Chemical and Pharmaceutical Industries. Blackwell Publishing Ltd; 2005. p 134–169.

[28] Schustera KC, Mertensb F, Gapesa JR. FTIR spectroscopy applied to bacterial cells as a novel method for monitoring complex biotechnological processes. *Vib Spectrosc* 1999;19: 467–477.

[29] Mourant JR, Yamada YR, Carpenter S, Dominique LR, Freyer JP. FTIR spectroscopy demonstrates biochemical differences in mammalian cell cultures at different growth stages. *Biophys J* 2003;85:1938–1947.

[30] Perin V, Goicoechea M. Monitoring substrate and products in a bioprocess with FTIR spectroscopy coupled to artificial neural networks enhanced with a genetic-algorithm-based method for wavelength selection. *Talanta* 2006;68:1005–1012.

[31] Ulber R, Frerichs JG, Beutel S. Review: optical sensor systems for bioprocess monitoring. *Anal Bioanal Chem* 2003;376:342–348.

[32] Fahrner RA, Blank GS. Real-time control of antibody loading during protein A affinity chromatography using an on-line assay. *J Chromatogr A* 1999;849:191–196.

[33] Fahrner RA, Lester PM, Blank GS, Reifsnyder DH. Real-time control of purified product collection during chromatography of recombinant human insulin-like growth factor-I using an on-line assay. *J Chromatogr A* 1998;827:37–43.

[34] Wold S. Cross-validatory estimation of the number of the principal components in factor and principal components models. *Technometrics* 1978;20:397–405.

[35] Geladi P, Kowalski B. Partial least squares regression: a tutorial. *Anal Chim Acta* 1986;185: 1–17.

[36] Lee W, Pasumarthy M, Undey C, Kundu S, Stout J. Assessment of Production-Scale Protein Purification Column Packing Through Transition Analysis. RXII: Recovery of Biological Products XII, 2006.

[37] Larson TM, Davis J, Lam H, Cacia J. Use of process data to assess chromatographic performance in production-scale protein purification columns. *Biotechnol Prog* 2003;19: 485–492.

[38] Molony M. Assessing process and quality attributes with various analytical technologies: Biogen Idec's PAT initiative. Oral Presentation, Well Characterized Biopharmaceuticals Meeting, Washington DC, 2007.

[39] Cooley R. Utilizing PAT to monitor and control bulk biotech processes. Oral Presentation, University of Michigan Pharmaceutical Engineering Seminars, March 4, 2003.

[40] Natarajan V, Purdom G. Advances in process chromatography gradient elution using binary linear gradients. *J Chromatogr A* 2001;908:163–167.

[41] Wu C-H, Scampavia L, Ruzicka J, Zamost B. Micro sequential injection: fermentation monitoring of ammonia, glycerol, glucose, and free iron using the novel lab-on-valve system. *Analyst* 2001;126:291–297.

[42] Cooley R. On-line process control: automating the control of process-scale purification columns using on-line liquid chromatography. *BioPharm Int* February, 2007.

[43] Rathore AS, Sharma A, Chilin D. Applying process analytical technology to biotech unit. *BioPharm Int* August, 2006, 48–57.

[44] Mattes, et al Analysis of Residual Moisture in a Lyophilized Pharmaceutical Product by Near-Infrared Spectroscopy. August Foss Application Note PH-AN-704, 2006.

[45] Almeida C, Calado C, Bernardino S, Cabral J, Fonseca L. A flow injection analysis system for on-line monitoring of cutinase activity at outlet of an expanded bed adsorption column almost in real time. *J Chem Technol Biotechnol* 2006;81:1678–1684.

[46] Reicht G. Near-infrared spectroscopy and imaging: basic principles and pharmaceutical applications. *Adv Drug Deliv Rev* 2005;57:1109–1143.

[47] Tang X, Nail S, Pikal M. Evaluation of manometric temperature measurement, a process analytical technology tool for freeze-drying: Part I, product temperature measurement. *AAPS PharmSciTech* 2006;7(1):E95–E103.

[48] Schneid S, Gieseler H, Kessler W, Pikal MJ. Process analytical technology in freeze drying: accuracy of mass balance determination using tunable diode laser absorption spectroscopy (TDLAS). Poster Presentation at AAPS Annual Meeting and Exposition, San Antonio (TX), USA, Oct. 29–Nov. 2, 2006.

[49] Reich G. Noninvasive and simultaneous monitoring of chemical and physical properties: an analytical approach to lyophilized proteins, sugars and protein/sugar mixtures. Proceedings of 3rd World Meeting APV/APGI, Berlin, 3–6 April, 2000. p 279–280.

[50] Kalkert K, Reich G. Near infrared spectroscopic evaluation of stress-induced structural changes of proteins in solid and solution state. Proceedings of 3rd World Meeting APV/APGI, Berlin, 3–6 April, 2000. p 335–336.

[51] Druy N. From laboratory technique to process gas sensor: the maturation of tunable diode laser absorption spectroscopy. *Spectroscopy* 2006;1–4.

[52] Voute N, Dooley T, Peron G, Lee E. Disposable technology for controlled freeze–thaw of biopharmaceuticals at manufacturing scale. *BioProcess Int* 2004.

[53] Lanan M, Kiistala M. Monitoring of minimum essential medium for mammalian cell culture. Oral Presentation, CASSS PAT Meeting; 2006.

[54] Lanan M, Kiistala M, Houde D, Donegan M. Flow-Injection ESI MS of Cell-Free Bioreactor Supernatants. Oral Presentation, IFPAC, 2006.

[55] Ramaswamy S, Greenberg AR, Peterson M. Ultrasonic detection of defects in highly porous polymeric membranes. *Insight* 2007;49:651–656.

[56] Schubert T, Breuer U, Harms H, Maskow T. Calorimetric bioprocess monitoring by small modifications to a standard bench-scale bioreactor. *J Biotechnol* 2007;130:24–31.

[57] Stokelman A, Slaff G, Zarur A, Rogers S, Low D, Seewoester T. High throughput cell culture experimentation with bioprocessors, SimCell platform. Biochemical Engineering XIV, Harrison Hot Springs, Canada, July, 2005. p 10–14.

[58] Nomikos P, MacGregor JF. Multivariate SPC charts for monitoring batch processes. *Technometrics* 1995;37:41–59.

[59] Tracy ND, Young JC, Mason RL. Multivariate control charts for individual observations. *J Qual Technol* 1992;24:88–95.

[60] Wold S, Geladi P, Esbensen K, Ohman J. Multi-way principal component and PLS analysis. *J Chemom* 1987;1:41–56.

[61] Nomikos P, MacGregor JF. Monitoring batch processes using multiway principal components analysis. *AIChE J* 1994;40:1361–1375.

[62] Wold S, Kettaneh N, Friden H, Holmberg A. Modelling and diagnostics of batch processes and analogous kinetic experiments. *Chemom Intell Lab Syst* 1998;44:331–340.

[63] Kourti T, MacGregor JF. Process analysis, monitoring and diagnosis using multivariate projection methods. *Chemom Intell Lab Syst* 1995;28:3–21.

[64] Cinar A, Parulekar SJ, Undey C, Birol G. Batch fermentation: modeling. In: Monitoring and Control. New York: Marcel Dekker; 2003.

[65] Undey C, Ertunc S, Cinar A. Online batch/fed-batch process performance monitoring, quality prediction, and variable-contribution analysis for diagnosis. *Ind Chem Eng Res* 2003;42:4645–4658.

[66] Wold S, Cheney J, Kettaneh N, McCready C. The chemometric analysis of point and dynamic data in pharmaceutical and biotech production (PAT): Some objectives and approaches. *Chemom Intell Lab Syst* 2006;84:159–163.

[67] Undey C. Are we there yet? An Industrial perspective of evolution from postmortem data analysis towards real-time multivariate monitoring and control of biologics manufacturing processes. IBC's 5th Annual Process Quality Forum: Process Characterization and Control, June 2–4, Cambridge, MA; 2008.

[68] Hofmann M. A novel technology for packing and unpacking pilot and production scale columns. *J Chromatogr A* 1998;796:75–80.

[69] Moscariello J, Purdom G, Coffman J, Root TW, Lightfoot EN. Characterizing the performance of industrial-scale columns. *J Chromatogr A* 2001;908:131–141.

[70] Conley L, Thommes J. Removal of process related impurities: in-process testing, validation and chromatogram review. IBC's 8th International Conference on Process Validation for Biologicals, March 7–8, San Diego, CA; 2005.

[71] Rudd D. The use of acoustic monitoring for the control and scale-up of a tablet process. *J Proc Anal Tech* 2001;1(2):8–11.

[72] Halstensen M, De Bakker P, Esbensen KH. Acoustic chemometric monitoring of an industrial granulation production process: a PAT feasibility study. *Chemom Intell Lab Syst* 2006; 84: 88–97.

[73] Naelapaa K, Veski P, Pedersen JG, Anov D, Jorgensen P, Kristensen HG, Bertelsen P. Acoustic monitoring of a fluidized bed coating process. *Int J Pharm* 2007;332(1–2):90–97.

13

EVOLUTION AND INTEGRATION OF QUALITY BY DESIGN AND PROCESS ANALYTICAL TECHNOLOGY

Duncan Low and Joseph Phillips

13.1 INTRODUCTION

Biotechnology has delivered some of the most significant advances in medical technology over the past 25 years. From the licensing of the first recombinant proteins in 1981, through the development of monoclonal antibodies, fusion proteins, gene therapy, transgenic animals and plants, stem cell technology, and now the creation and selection of completely novel molecules through techniques such as molecular evolution and phage display, the degree of innovation has been remarkable. More importantly, the contributions to patient survival, quality of life, and eventual cure has been, and continues to be, a source of great pride and motivation to our industry.

In contrast, the extent to which the pharmaceutical industry has incorporated advances and improvements in manufacturing technology has been considerably less adventurous. It seems that any degree of innovation and creativity can be applied to the creation of new candidate molecules, but once those molecules are recognized as having potential, without any deleterious side effects, any change to the molecule or the process by which it is made is viewed as a regulatory hurdle and an opportunity for failure rather than an opportunity for continued improvement and refinement.

Quality by Design for Biopharmaceuticals, Edited by A. S. Rathore and R. Mhatre
Copyright © 2009 John Wiley & Sons, Inc.

Changes to processes are avoided, variability has been viewed as inherent and controlled by testing and rejection rather than by adaptive processing, and quality and process performance suffer as a result. Admittedly, much of this stems from legitimate concerns for patient safety on the sides of both manufacturers and regulatory authorities, but in the end the greatest benefit to patients is perceived to be an industry, which is able to manufacture high-quality products in an efficient manner. These efficiencies would enable it to focus its resources on the continued advancement of new modalities rather than wasting them on ineffective processes, particularly when compared to manufacturing from other industries such as food, microelectronics, and petrochemicals.

Given that the most frequently quoted reason for reluctance to introduce innovations was the regulatory environment, the FDA established several initiatives, including the process analytical technology (PAT) initiative, with the objective of removing obstacles to innovations which could lead to superior quality and manufacturing efficiencies. PAT is essentially an approach to secure quality and process consistency through a combination of process understanding and real-time responsive control. It begins with initial process design, allows dynamic processing, minimizes waste, and supports continuous improvement throughout the product life cycle.

More recently, additional communications from the FDA have focused on the broader area of pharmaceutical cGMPs for the twenty-first century, and ICH guidelines have further developed the concepts of Quality by Design (QbD) and design space. The original FDA PAT team has been restructured, leaving some to wonder if the original commitment to PAT and the promise of regulatory flexibility still stand.

In this chapter, we shall review briefly the development of PAT and QbD in FDA, EMEA, and ICH guidelines, describe how industry initiatives in developing standards and guidelines are intended to support and elaborate on the guidelines, and discuss approaches for the implementation of PAT in securing quality, reducing variability, and supporting continuous improvement in biotechnology processes.

13.2   EVOLUTION OF PAT AND QUALITY BY DESIGN (QbD): EMERGING GUIDELINES AND STANDARDS

To understand how the emphasis may have seemed to shift away from PAT, it is helpful to consider the broad time line of events as the agency developed its program to reshape the way in which the pharmaceutical industry approached lifecycle management of manufacturing science and technology (Fig. 13.1), see also Chapter 2.

In summary, although the initial dialogue seemed to focus mostly on PAT [1, 2], as the agency produced additional documents [3, 4] it became apparent that although the agency continued to state its support for PAT, the PAT initiative was part of a grander scheme to address quality, and what was perceived as stagnation in the delivery of new therapies andin manufacturing technology and innovation. In 2005, the agency initiated the Officeof New Drug Quality Assessment (ONDQA) CMC Pilot Program to revamp the way in which filings were submitted and reviewed. The new Pharmaceutical Quality Assessment System (PQAS) placed greater emphasis on scientific knowledge and

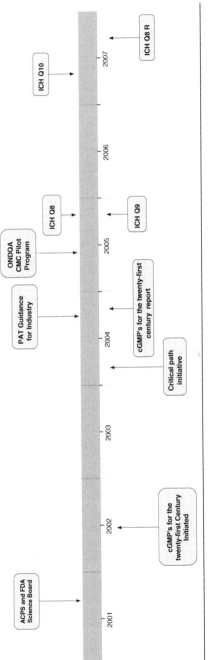

Figure 13.1. FDA initiatives in QBD and PAT (with thanks to Dr Moheb Nasr).

understanding of the product and process by applying Quality by Design principles, and built on the team-based integrated review and assessment process that was developed for PAT; companies were encouraged to file applications under the new system with an agreement that if the appropriate level of process and product understanding was demonstrated, then the companies would be free to make changes within the approved design space based on internal quality systems and GMP controls.

The greater flexibility available permits greater freedom to make improvements in the process and facilitates the transfer of processes from one site to another and even one manufacturer to another, provided the appropriate analytical tools and assays are available.

A major industry concern is that none of these changes should trigger differences in filing approaches from one region to another. The International Conference on Harmonization (ICH) has sought to avoid the development of separate approaches through guidelines for quality, safety, efficacy, and multidisciplinary topics. The ICH Guidelines Q8, Pharmaceutical Development, which describes the concept of design space, and Q9, which provides guidelines for risk management, were both finalized in 2005, with an annex to Q8 (R1) expanding on the description of design space (albeit written in the context of small-molecule drugs) becoming available, but not yet finalized, in 2007. A draft became available for Q10 in 2007, which addresses pharmaceutical quality systems [5–8].

The EMEA PAT team was formed in late 2003 to review the implications of the PAT initiative and ensure that the European framework and regulatory authorities were adequately prepared to conduct effective evaluations. The EMEA inspections Web site for process analytical technology states that "Quality by Design is an established concept in Europe" [9]. In general the EMEA position is that the current regulatory framework in Europe is open to the implementation of PAT in marketing authorization applications through existing guidance for pharmaceutical development and parametric release. The EMEA published a reflection paper in 2006 [9], and there is a document of questions and answers on their Web site [10].

During the development of this framework, it was recognized that the PAT guidance would be general and would not provide specifics, which would be developed instead in the form of best practices. The FDA encouraged the pharmaceutical industry to take an active role in drafting these practices in the form of consensus standards, which led to the formation of the ASTM E55 Committee on Manufacture of Pharmaceutical Products in 2003, shifting the onus for continued development of the concepts to the manufacturers. The original aim was to develop standards to support the implementation of PAT, but the committee subsequently broadened its scope to all aspects of manufacture to include the "development of standardized nomenclature and definitions of terms, recommended practices, guides, test methods, specifications, and performance standards for the manufacture of pharmaceutical products" [11]. Quality and consistency are additional goals, as are transparency and cost effectiveness.

Standards can be general recommendations, such as guides, or can be increasingly specific recommendations, such as practices and specifications. They are developed in a consensus process, which requires that they are open (all can participate), balanced (i.e., no interest group can dominate), they follow due process (all points of view are addressed), and they are relevant and coherent (such that they do not duplicate existing standards) [12]. The E55 committee includes members from both brand name and generic pharmaceutical manufacturers, biopharmaceutical and biotechnology companies, regulators, consultants, suppliers, and academia, and represents a broad base of interests and expertise.

The use of standards is voluntary but ensures compliance if they are followed. US government agencies, including the FDA, are mandated by law, via the National Technology Transfer and Advancement Act of 1995 [13], to use voluntary consensus standards in lieu of government-unique standards, except where inconsistent with law or otherwise impractical [2]. The agency sees standards as "an optional way to describe how a certain function will be implemented to meet requirements. FDA describes what has to be done to meet the requirements and the companies may use standards to describe how it is done" [14]. The benefit of this approach is that companies can have a degree of confidence in the acceptability of certain standards or practices, and regulators have the benefit of participation in, and oversight of, the rigor behind the recommended practice. To date, four standards have been approved (including new approaches to equipment verification, streamlining IQ/OQ/PQ; and continuous verification), and over a dozen of others are in various stages of development, as given in Table 13.1.

The standards are valuable tools for the development and implementation of modern manufacturing practices, ensuring quality and capturing value. For example, the systems verification standard, E2500, offers a modern approach to process verification as applied to manufacturing systems and equipment, redefining the role of quality and streamlining many of the repetitive and nonvalue added tasks in traditional commissioning and qualification testing [15]. The standards for risk management and process understanding are close to approval, since any discussion over "regulatory flexibility" will be commensurate with risk, and in proportion to the level of process understanding that has been demonstrated, it will be helpful to share a common understanding of what those terms mean with regulators.

ISPE has formed a PAT Community of Practice that serves as a forum to exchange ideas and share approaches with other practitioners. The group organizes workshops and meetings for training in PAT practices and issues, such as data management, regulatory approaches, and business case for PAT. ISPE produces "Baseline Guides" to assist members in meeting regulatory expectations for facilities and manufacturing equipment and is in the process of revising the guide for Commissioning and Qualification to bring it into alignment with the ASTM E2500 standard discussed earlier [15]. Education around PAT was also seen as a priority, which led to the formation of the PAT Community of Practice by ISPE. Furthermore, the Product Quality Lifecycle Implementation Initiative was launched in 2007 to provide the technical framework for implementation of QbD and to enable adoption of ICH Q8, Q9, and Q10 documents. Again, regulators are engaged with ISPE communities of practice, which will ensure the continued development of PAT and QbD approaches; none of this detracts from the earlier commitment to PAT. The initial guidance was clear that the procedures would be "consistent with the basic tenet of quality by design" [2].

TABLE 13.1. Standards from the ASTM E55 Committee on Manufacture of Pharmaceutical Products

Approved Standards	Scope
E2474-06 standard practice for pharmaceutical process design by using process analytical technology	Development and implementation of process understanding
E2500-07 standard guide for specification, design, and verification of pharmaceutical and biopharmaceutical manufacturing systems and equipment	All systems with the potential to affect product quality and safety
E2503-07 standard practice for qualification of basket and paddle dissolution apparatus	Setup and calibration of dissolution apparatus.
E2537-08 standard guide for application of continuous quality verification to pharmaceutical and biopharmaceutical manufacturing	Validation of processes where performance is monitored continuously.
E2363-06a standard terminology relating to process analytical technology in the pharmaceutical industry	Defines terminology specific for PAT as used in the pharmaceutical industry
Standard practice for risk management as it impacts the design and development of processes for pharmaceutical manufacture	Risk management and assessment for processes using PAT principles.
Standard practice for process understanding related to pharmaceutical manufacture and control	Builds on understanding as described on the PAT Guidance
Standard guide for the application of continuous processing technology to the manufacture of pharmaceutical products	Identifies principles of continuous processing for pharmaceuticals
Standard practice for process sampling	Sampling considerations for PAT
Standard practice for qualification of PAT systems	Qualification of systems using PAT principles
Standard guidance for multivariate analysis related to process analytical technology	Development and use of MVDA models and controls
Standard guidance for identification of critical attributes of raw materials in pharmaceutical industry	Applies knowledge and understanding to selection of materials
Guide for validation of PAT methods	Applies to in-line, online, and near-line spectroscopic methods.
Online total organic carbon (TOC) method validation in pharmaceutical waters	Validation of online analyzers
Guide for science-based and risk-based cleaning process development and validation	PAT for cleaning processes and cleaning validation

13.3 PROCESS ANALYTICAL TECHNOLOGY (PAT)

PAT requires the use of online, in-line, or at-line sensors to monitor continuous or semicontinuous operations, and it uses the information to make real-time or near real-time control decisions about critical process parameters and/or product quality. It was originally developed for telemetry or remote sensing via wire, cable, telephone, or wireless technology. Telemetry has been used by NASA for over 45 years for a wide range of applications, monitoring both machines and men [16].

The development of microprocessor-based programmable logic controllers during the 1970s was an important step for PAT as it now allowed control schemes using feedback loops to monitor and respond to conditions based on the received signal. Distributed control system (DCS) and supervisory control and data acquisition (SCADA) system led to the integrated sensor and control systems now widespread in process industries such as petrochemical, pulp, paper, and nuclear power plants. Process analytical instrumentation was introduced in the 1980s and total organic carbon analyzers and conductivity/resistivity sensors became standards in the semiconductor industry 10–15 years before their acceptance in the pharmaceutical industry [16].

The pharmaceutical industry has been relatively slow to adopt this approach but the last 5 years have seen a dramatic increase in interest, with multiple conferences and magazine articles addressing the adoption of both PAT and QbD appearing in the United States and overseas. There are multiple reasons for the increase in interest, beginning with the desire for improvements in manufacturing technology and quality initiated by regulators, more recently shifting to the increasing pressure on manufacturers to pursue opportunities for continuous improvements in process economics and reductions in the cost of quality. A 2006 survey suggested that the pharmaceutical industry may be wasting as much as $50 billion out of $200 billion spent annually on drug manufacturing [17].

The pharmaceutical industry is responding by engaging lean manufacturing practices, which seek to optimize production processes by the elimination of wasteful practices and by streamlining processes through a focus on flow. Traditionally the industry has used batch and queue processes with multiple hold points. PAT supports lean manufacturing both directly and indirectly by focusing on several of the seven wastes of lean; specifically defects, waiting, overprocessing, and inventory (the remainder being overproduction, transport, and unnecessary motion). It has been proposed that PAT can improve quality management and reduce costs in quality appraisal and failures, and that investment in prevention, such as QbD, continuous improvement, and PAT-enabled controls reduce the total cost of quality by improving process capability [18]. This study proposed that the major economic impact of PAT lay in its ability to reduce queues by enabling real-time release of process intermediates and products, thereby reducing cycle times and working capital (carrying costs of 20–25% were assumed for inventory). Companies from biotech, branded pharmaceutical, and nonpharmaceutical industries were benchmarked, and it is of note that biotech had the poorest turnover rates of the group, suggesting that there are major opportunities for improvement. The extent to which real-time release may be achievable for biotech products, where sterility is a major issue, is debatable, but there is little doubt that in-process holds for intermediates can be shortened with greater adoption of PAT principles.

In addition to the ASTM and ISPE groups mentioned above, several other groups offer opportunities for interaction with regulators and the exchange of best practices. The Parenteral Drug Association (PDA) has a quality and regulatory department that interacts with FDA, EMEA, WHO, ICH, USP, and numerous other regulatory bodies around the world, and it also notifies members for updates in regulatory initiatives. The PDA also develops technology reports and guidance documents where they perceive that gaps exist, and it also arranges conferences with a focus on regulatory, manufacturing, and science and technology topics around the world. The International Foundation for Process Analytical Chemistry (IFPAC) runs annual meetings to share information on PAT across a broad industry base. Initially dominated by large chemical and petrochemical manufacturers, IFPAC is increasing its focus on the pharmaceutical and biotechnology industries, with the FDA functioning as one of the major sponsors and providing three of the conference co-chairs over the past 4 years. It is unusual as one of the few conferences supports cross-industry learning in an increasingly specialized world. The Pharmaceutical Process Analytics Roundtable (PPAR) is an informal discussion group that meets annually to discuss PAT-related topics in the pharmaceutical and biotechnology industries.

13.3.1 PAT Tools

As described in the PAT Guidance, PAT tools fall into four main areas: process analyzers, process control tools, multivariate tools for design, data acquisition, and analysis, and continuous improvement and knowledge management tools [2].

There are a host of *analyzers* suitable for gathering information from biotech processes, ranging from standard sensors such as pH, conductivity, and temperature, through more sophisticated spectroscopic techniques, to sophisticated online analyzers. UV is universally used to detect and track protein products and to make processing decisions as to where to collect product or divert process streams to waste, but it is not capable of discriminating between product and closely related contaminants. Near infrared (NIR) is widely used in small-molecule processes and is useful for testing the identity of incoming materials and with the potential to yield actionable information about the condition of upstream processes [19]. Online high-performance liquid chromatography (HPLC) has been used to determine peak quality in purification processes [20, 21], but the problem downstream lies in shortening analysis times to seconds rather than minutes. There is a significant need for developments in sensor technology, specifically to cope with biological molecules and the relationships between structural integrity/aggregation and biological properties. This gap was highlighted at a recent workshop held at IFPAC 2008 and co-chaired by the FDA, NIH, NIST, and MIT. Obvious candidates would be methods for rapid determination of glycosylation, the presence of aggregates, and contaminants such as endotoxins [22, 23].

Multivariate data analysis (MVDA) is an essential tool. MVDA applies statistical tools such as principal component analysis (PCA) and partial least squares (PLS) to complex data sets to study their multidimensional nature and the interdependence of multiple material attributes and process parameters [24]. These tools can be readily applied to upstream applications where there is a multitude of input and process parameters that can be studied [25] and whose interactions may need to be defined.

In purification, correlation between material attributes (e.g., buffers and resin properties), process parameters, and outputs can lead to the development of models that give a better indication of packing integrity and process performance [26]. Care should be taken with MVDA approaches to establish a causative relationship between observed trends and a mechanistic understanding of the process; see also Section 13.5.1.

Finally, a number of software tools to support MVDA and to enable *continuous improvement* and *knowledge management* are becoming available, but equally as important, the regulatory mechanisms must be in place to support changes when changes bring such improvements within established design space. As stated earlier, the regulators are not planning to develop additional guidance documents but are instead looking to the industry to develop and implement best practices through consensus processes. We must continue to look to the groups and professional societies detailed earlier for best practices. How the tools are deployed is discussed in a later section and in Chapter 12.

13.4 QUALITY BY DESIGN

The traditional approach to biopharmaceutical manufacturing processes has been that the "process is the product." This can result in regulatory filings that are descriptive of the process, starting at the beginning and detailing specific activities at each point, rather like a recipe or itinerary. Such submissions can be rich in descriptive detail in terms of what is done at each step of the process, but lack an equally detailed explanation as to why specific steps were chosen and what impact the chosen process parameters had on process performance or product quality attributes and how they interact with each other. The QbD approach however places greater emphasis on final product quality attributes and is more concerned with the outcome than with the process by which it was arrived at. A demonstration of *product* and *process understanding* is desired. The QbD approach, as outlined in ICH Q8 (R1) [6], begins with *product design* (Fig. 13.2). The target product profile (TPP) is first ascertained, with the primary focus being on safety and efficacy. Additional attributes that may impact manufacturability, stability, and so on (such as post-translational modifications, known "problematic" sequences) should also be considered, see Chapter 2. This is a particularly rich opportunity area for biotechnology products, since by nature they are considerably more complex than small molecules. A single modality such as a monoclonal antibody can provide sufficient diversity to address a multitude of targets as either an agonist or an antagonist, yet still be sufficiently similar such that common platforms of production and purification approaches can be used [27]. This allows relevant process information from multiple projects to be gathered and compared to enhance process understanding.

Critical quality attributes (CQAs) are then derived from the TPP and are linked to material attributes and process parameters based on their impact on safety and efficacy, and a *risk assessment* is performed as a function of all unit operations and material inputs. The purpose of the risk assessment is to focus attention on those variables and parameters which are most likely to have an impact on final product quality, such that an appropriate level of understanding is developed and appropriate monitoring and control strategies can be designed and implemented. Additional parameters may be further designated

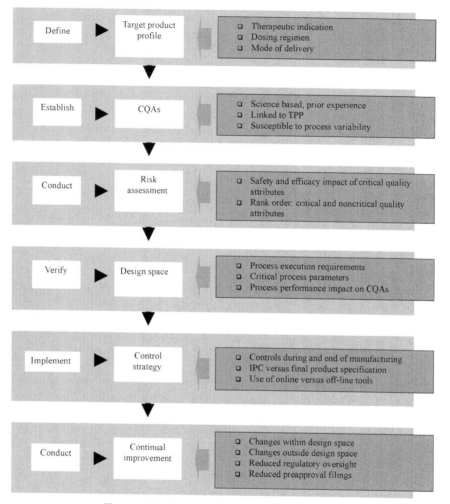

Figure 13.2. The QbD approach (from ICH Q8R1).

"key" or "nonkey" if they impact process performance (e.g., yield) without impacting quality, or if they have no impact on either in the range studied.

The linkage between inputs and quality attributes is described in the *design space*, which is a multidimensional description of the conditions (input variables and process parameters), which result in a product with the desired CQAs being produced. It can be based on a unit operation or the overall process. Design space can be defined in scale and equipment independent terms, which can facilitate technology transfers between scales or even sites as the project or commercial process matures.

The next step is to design and implement a *control strategy*. Here is where PAT plays a major role, and it can be used independently or together with additional strategies, including end-product testing. PAT approaches often require continuous or at least frequent monitoring and allow the process to respond to any variation in incoming attributes by modifying conditions through either feedback or feed-forward control such that the desired outcome is still achieved. This can be accomplished as a response to either a single or, in more advanced control situations, multiple inputs, and results in the control of single or multiple parameters. Finally, product quality should be managed throughout the *product life cycle*, and as new approaches are developed or new knowledge and understanding are acquired, quality should undergo a process of *continual improvement*, which is both simplified by the preceding practices and assumed as a quality management principle.

13.4.1 Combining QbD and PAT

The scope of the different ICH guidances, and how they relate to each other, is shown diagrammatically in Fig. 13.3, along with where QbD and PAT are primarily applicable. QbD and PAT are strongly interdependent, but it is also the case that certain aspects of QbD do not require PAT for implementation, such as product or formulation design, and equally, there are opportunities for PAT in areas outside direct process control, such as monitoring equipment for continued fitness for use (e.g., environmental monitoring) and cleaning. However, when it comes to direct control of process parameters and product attributes, QbD serves an essential part by first defining the required design space. Once defined, PAT is the execution tool that keeps the process in the desired space.

Supporting QbD principles with PAT allows the development of a dynamic process that can respond to input variability. It follows that such a process should be better able to respond to changes and facilitate the incorporation of improvements in process technology, be it the changes in equipment or in critical processing aids such as resins and filters [28].

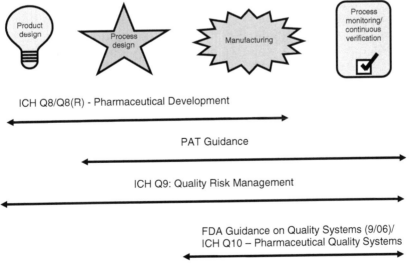

Figure 13.3. The scope of recent guidances (with thanks to Dr. Moheb Nasr).

13.5   IMPLEMENTING QbD AND PAT

If a manufacturer is to implement a QbD business process and reap the full benefits of consistent product quality, responsive processing and the ability to incorporate improvements on a continuous basis within their own quality system, the approach to developing the process will probably differ from the traditional approach in a number of ways. The principle of these would be the amount of data that is gathered and the way in which it is presented. Since the basis for QbD is a well-defined design space, there will be a greater degree of experimentation required upfront to fully explore the relationships between input materials and process parameters. Second, since the focus has shifted from a detailed description of the process to its outcomes and a demonstration of process understanding, there should be more emphasis on the knowledge gained during development and an explanation of the purpose behind each step (or series of steps) and how the desired end points are achieved. It is also necessary to establish a common basis for what is meant by "process understanding," see later. Detailed descriptions of control strategies will replace detailed descriptions of processes and quality will be managed over the life cycle of the product through continuous improvement. A comparison between traditional and QbD/PAT processes is given in Table 13.2.

TABLE 13.2.   A Comparison of Traditional and QbD/PAT Processes

Aspects	Traditional	QbD
Product design	Screening, the best candidate wins	TPP developed though broader use of prior knowledge, consideration of manufacturability
Pharmaceutical development	Empirical; typically univariate experiments	Systematic; multivariate experiments, PAT tools used
Manufacturing process	Fixed	Adjustable within design space; opportunities for innovation (PAT)
Process control	In-process testing for go/no-go; offline analysis	PAT tools utilized for feedback and feed-forward controls
Product specification	Primary means of control; based on batch data at time of submission	Part of the overall quality control strategy; based on desired product performance (safety and efficacy)
Control strategy	Mainly by intermediate and end-product testing	Risk-based; controls shifted upstream; reducing product variability; real-time release
Life cycle management	Reactive to problems and OOS; postapproval changes needed	Continual improvement facilitated

Source: Modified from ICH Q8 R1 and with thanks to Moheb Nasr.

13.5.1 Process Understanding

Process understanding, or rather an understanding of the interdependencies between the inputs and process parameters, the target product profile and quality attributes, is fundamental to both QbD and PAT. A Standard Guide to Process Understanding is being developed by the ASTM E55.01 subcommittee to provide a better definition and clearer common vocabulary for this important topic [29].

A process is considered well understood when "all critical sources of variability are identified and explained, the effect on quality attributes can be accurately and reliably predicted based on the inputs to the process, and the process capability for critical quality attributes meets the required acceptance levels." The standard recognizes that learning continues throughout the life cycle of a product and that additional knowledge will become available as additional experience is gathered or as new PAT tools are brought to bear. Typically, process information is first gathered through the appropriate small-scale models and experience with similar processes, but knowledge and understanding of the process develops as experience is gained through various scales and stages and modifications through and beyond commercialization.

Process knowledge exists in various states, as indicated in Table 13.3.

The most basic level of understanding, *descriptive knowledge*, is simply a description of what is done and is usually displayed as high-level flow charts and descriptive text. In this case the focus is on compliance and any excursions take the process into unknown territory. *Correlative knowledge* is achieved when the interrelationships between variables are known but without understanding the underlying mechanisms. Correlative knowledge can be improved through, for example, risk analysis. *Causal knowledge* is achieved when there is significant scientific knowledge on the underlying causes of the interrelationships between variables. There is a level of awareness to the criticality of variables and approaches to their control through PAT tools. *Mechanistic knowledge* is based on understanding how causal relationships occur. It is typically achieved through multivariate analysis and executed through multivariate statistical controls. At this stage

TABLE 13.3. Categories of Process Knowledge from Lowest to Highest State of Understanding

Descriptive knowledge	What is occurring	Derived from observation and reflects basic facts, focus on compliance
Correlative knowledge	What is correlated to what	Correlation between inputs and outputs, may be univariate, FMEA
Causal knowledge	What causes what	Significant scientific knowledge established, critical variables understood, PAT controls
Mechanistic knowledge	How	Multivariate tools and controls employed, solid basis for scale-up and tech transfer
First principles understanding	Why	Fundamental theoretical understanding, prediction of behavior, design in performance, and process innovation

the process is robust. The final level of knowledge, *first principles understanding*, reflects knowledge of the underlying "why" interrelationships occur, and as a result it is highly predictive. Insights at this level form a basis for process innovations and generalized to other process steps.

If the view is taken that the "process is the product," the result is that the process is described in terms of descriptive, and perhaps correlative, knowledge. The opportunity that may seem to be missed is that higher levels of knowledge and understanding are frequently not developed unless prompted by an excursion or a quality issue. Generally, most companies have gone far beyond this level, and there is far more data to support their products and processes; the shortcoming lies in documenting and demonstrating their understanding to the regulatory bodies, perhaps in order to avoid prompting questions or discussions that are not necessarily value added in terms of the impact on quality, but resulting in an understandable tendency on the part of the regulators to take a compliance-based approach to inspections and a restrictive view to change. Manufacturers who can demonstrate the highest levels of understanding, on the contrary, may enjoy the confidence of regulators and deploy process strategies with greater freedom to innovate.

13.5.2 Product Design

The area of product design is rich in opportunities for QbD approaches, but less so for PAT, since once the optimal molecule has been chosen that part of the process is fixed. The same can be said for the selection of clones and expression systems—there may be considerable experimentation to determine the best combinations of growth characteristics and product expression, but once clone selection is finalized that part of the design is locked down. PAT tools can be employed to investigate how different product candidates respond in scaled-down processes, and thus provide insight on manufacturability, an important selection criteria when other factors may be closely similar or equal. High-throughput screening tools are of considerable value in this area.

13.5.3 Process Design and Development

Developing a process that integrates PAT tools and principles and demonstrates science- and risk-based process understanding will give rise to an approach that deviates from the traditional filing strategy, which was focused on the process and resulted in largely descriptive submissions. The Standard Practice for Process Design using PAT was released in 2006 [30], and it describes eight basic practices that should be followed to ensure proper design, reduce variation, and meet the required acceptance levels for quality (Table 13.4). The approach is equally applicable to drug substance and drug product.

Given that the emphasis from the agency has been on a risk-based approach to quality, the first practice is *risk management through assessment and mitigation*, which is applicable throughout design. The objective is to reduce variability and ensure safety of the process through formal risk evaluation procedures. Risk assessment and control is the subject of a separate standard currently under development; clinical safety is addressed in clinical trials.

TABLE 13.4. Basic Practices for Process Design by Using PAT

Practice	Objectives
Risk assessment	Reduce variability, ensure safety
Continuous improvement	Capture benefits of increased understanding
Process fitness for purpose	Process capability and robustness
Intrinsic performance assessment	Built in monitoring systems, continuous measurement
Manufacturing strategy	Minimize risk
Data collection and formal experimental design	Study both simple and complex interactions
Multivariate tools	Develop sophisticated predictive models
Process analyzers	Capture process data
Process control	Based on high-level understanding and process performance

The second practice captures the dynamic nature of PAT approaches and addresses *continuous improvement*. Before starting process development, consideration will be given to knowledge about product structure and process performance from similar products and processes. Continued evaluation of design options is an iterative process that continues throughout the life cycle of the product. The original design space and the models used to establish process design should be revisited from time to time to see if they can be improved in the light of new data. Equally, as new analyzers permit fresh insight into process performance and product quality, the design space should be modified in light of the new information. The degree to which this can be done will be dependent on risk and the level of understanding.

Quality and risk assessments should continue to be applied at each step, to ensure that the third practice, *process fitness for purpose*, is followed. This ensures that the process is appropriately designed and has the required process capability to deliver the desired outputs and relates consistently. It includes product quality, process performance and process systems, and commercial viability.

The next practice is *intrinsic performance assessment*, and it differs from traditional approaches in significant ways. The first is that the process will be designed with intrinsic monitors and controls operating in harmony. Measurements will be made in-line or online rather than by sampling, monitoring will be continuous rather than on the basis of averages and will be focused on tracking product quality and process performance rather than meeting acceptable ranges. The process will be considered in its entirety, such that outputs from one operation are inputs to the next. The focus is on performance and capability rather than simply following instructions.

The *manufacturing strategy* should be designed to minimize risk. One aspect of this is the selection of scale and manufacturing technology, and as described elsewhere, the risks in transferring from one scale or even one site to another can be minimized if the process is designed to include scaleable or even scale-independent technologies. The process is described in dimensionless terms and controlled by process parameters and quality attributes rather than equipment parameters.

The next three practices, *data collection and formal experimental design*, the use of *multivariate tools*, and in-line, online, or at-line *process analyzers* are intended to optimize the quality and quantity of information about the process. They enable the development of sophisticated models based on detailed understanding of multiple interactions and enable the final practice, *process control*. When all critical quality attributes have been accomplished the process has reached its end point.

Because of their complexity, biotechnology processes typically have a greater degree of characterization and control than small molecule processes, and in many cases there are already a significant amount of relatively simple PAT controls around both upstream and downstream processes. Incoming raw materials have been studied to see how any variability may impact process and product quality and to develop suitable screening control initiatives and feed strategies [31]. Upstream raw materials used for cell culture or fermentation processes are a major source of variability, either because they are inherently complex (as in the case of hydrolysates) or, in the case of chemically defined media, because they contain a wide array of energy sources, growth factors, and trace nutrients. Multivariate tools are proving useful in determining the interactions between materials and process conditions [32], but because of the complexity of microbial or mammalian cell culture, it may be harder to claim higher level understanding of every detail of the process. PAT tools may be as useful in establishing what does *not* need to be measured and controlled in manufacturing as for what does.

Recovery and purification operations, on the contrary, are relatively simpler, and there are well-developed theoretical models for many of the typical unit operations for centrifugation, filtration, and chromatography. The problem in development is to establish the clinical significance of process- and product-derived contaminants, which may be monitored and controlled but whose significance will not be fully understood.

13.5.4 Manufacturing

If QbD, design space, and product knowledge are the instruments and objectives of process development, PAT analyzers, process understanding and controls, and continuous improvement are the tools and goals of the manufacturing engineer.

PAT controls are already in place for many biotech processes. Although QbD seeks to define a multidimensional design space, controls may be univariate or multivariate, depending on what is important for the process (e.g., monitoring to an end point, or more interactive control of a continuous process based on multiple parameters). PAT seeks to assure quality either directly, by monitoring and controlling product quality attributes during the process, or indirectly, by reducing variability and thereby providing a higher assurance of quality. PAT can also be used to provide continued assurance of fitness for purpose of process equipment and facilities, see Fig. 13.4. PAT can be applied at multiple levels, and just as QbD can be applied without use of PAT, so PAT can be applied to situations that are not specifically for the design space established by QbD.

The focus of QbD is largely on the design and development of the product to ensure efficacy and patient safety, and considers the product in terms of CQAs. Some parts of product design are fixed and cannot be impacted by PAT. Instead, PAT focuses on the

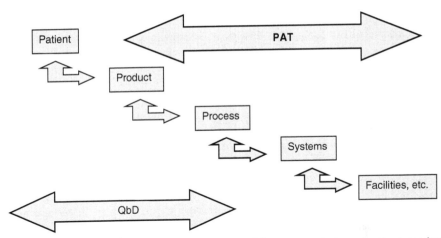

Figure 13.4. Relationships between different levels of the sequence between patient, product, process, and manufacturing components. PAT can be applied to areas that do not directly impact the process, such as environmental control.

interaction between the product, the process, and the materials from which it is made. Screening incoming materials for identity, contaminants, and possible adulterants ensures the integrity of the process from the outset and can now be performed at line with hand-held monitors. Monitoring and controlling pH and conductivity eliminate variability in raw materials or operator errors in the makeup of media and buffers. The fitness for use of pumps and processing equipment such as columns can be assessed through vibration monitors (on pumps) or by tracking transitional analysis (on columns). These parameters are not always directly within scope for QbD although they may enable the final outcome of the process. Equipment automation will be increasingly important as higher titers will result in greater dependence on the use of concentrates in downstream processing, requiring the use of automated in-line dilution systems.

There are several examples elsewhere in this book for where and how design space can be developed and PAT tools can be deployed, so we shall limit our discussion to two unit operations, one for drug substance and another for drug product, to give an interpretation as to how the tools can be used.

13.5.5 Drug Substance Purification

Chromatography is a critical purification tool in biotechnology. The development of a design space for a chromatography operation is described elsewhere in this book, so we shall focus on implementation aspects.

Assuming that there has been an initial risk assessment of the process, and that the basic conditions have been developed in such a way that CQA's have been identified and their dependency on process parameters is at least partially understood, what is the role of the manufacturing engineer in applying PAT principles to ensure the process remains within the design space and delivers product of desired quality? What opportunities are there for further refinement of the process and for continuous improvement through reduction of variability and elimination of waste? Applying the principles discussed in the previous section, we begin again with a risk assessment, this time focusing on the manufacturing operation rather than the design space. Inputs are typically organized into six elements in an Ishikawa diagram: people, equipment, measurement, method, materials, and environment (Fig. 13.5).

People as a source of variability can be mitigated by training, the availability of SOPs, and where appropriate, eliminated by the use of automation. Column packing in particular can be operator-dependent; conditions should be standardized as far as possible. Operator and scale dependence of pressure-flow curves can be minimized by ensuring slurry conditions are determined consistently and reproducibly [33]. Differences can still arise on the basis of resin properties (materials—resin density, particle size, and charge/hydrophobicity) and column designs (equipment—primarily flow distribution systems). Environmental conditions can have an impact on the separation, in particular temperature, either directly or indirectly. Hydrophobic interaction separations are prone to perturbation by temperature, but other effects should be considered such as bringing buffers or sample from cold storage onto a column at room temperature, and the activity of enzymes present as impurities. Materials and equipment have a considerable impact on the process, even if conditions are well defined from the development of design space (for example the selection of ligand, ion, and counter-ion in an ion exchange separation), see Fig. 13.6. Resins can show lot-to-lot variability in multiple parameters, such as ligand capacity, particle size and particle size distribution, and pore size and pore size distribution. This variability can contribute to significant differences in retention times and/or in elution profile [34], which would be difficult to accommodate in manufacture if tanks (for example) are not large enough to cope with the increased pool volumes. Buffers can also be a source of variability depending on the supplier and the grade of material chosen, and may even have different conductivities, despite being at the same molarity [35]. Much of this variability can be compensated for in the way the method itself is run.

In bind and elute methods, elution is frequently accomplished by running a gradient, which can be either step or linear in form. Step gradients have the benefit of being simpler to produce, and peaks are typically less dilute, but they are less discriminating since all of the pool is retained as product and there is less distinction between the leading and trailing edges of peaks. The specific performance of a given step will depend on the step conditions chosen and the criticality of the separation to be achieved. Linear gradients may be more likely to pickup differences in levels of minor product variants and closely related contaminants.

Linear gradients are a situation where, at least to some extent, "variability is managed by the process"—any variability in ion exchange capacity, and buffer conductivity (either from buffer source or from errors in buffer makeup) is compensated for by the use of a linear gradient. Retention times may vary from run to run but are in fact consistent if the process is defined in terms of critical process conditions such as pH, conductivity, resin capacity, flow rate, and other parameters found to be relevant during experimental development.

Finally there may be differences in equipment design, especially between one scale and the next (e.g., system dead volume, valve arrangement, and column distribution flow

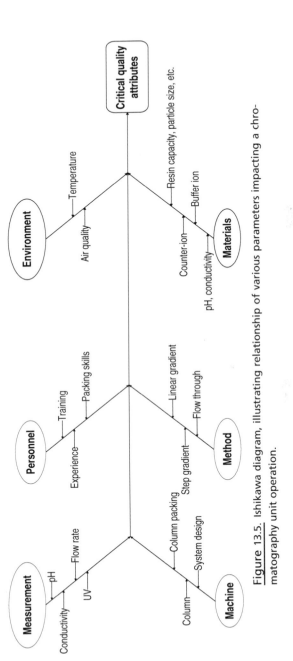

Figure 13.5. Ishikawa diagram, illustrating relationship of various parameters impacting a chromatography unit operation.

system), which can result in differences in gradient shape, which will further impact the process (Fig. 13.6b). Again, if the process is defined in the appropriate terms, it should be simple to justify modifications to operating conditions to ensure that the correct combination of process parameters, corresponding to the design space for that step, is delivered, see Fig. 13.6c.

During operation the system should be monitored for continued fitness of purpose. Packing integrity may fail with time, this can be mitigated by tracking performance using transition analysis of internal markers, which has been shown to be a useful predictor of potential failure [36]. As additional experience is gained with the separation, PAT tools can be used to fine tune performance, for example by modifying the steepness of the gradient or varying flow rates to shorten elution times, improve yields, increase loading, improve quality, or decrease buffer consumption, provided of course that attention is paid both to the behavior of the product of interest and the impurities and contaminants such that quality is not compromised. Again, this is a science- and risk-based assessment based on an understanding of differences in properties between the different molecules (e.g., pI and titration curves in the case of an ion exchange separation).

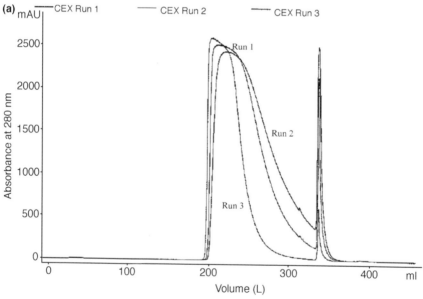

Figure 13.6. A variability introduced by materials and equipment. (a) Variation in retention time and pool volume resulting from differences in buffer conductivity. The three runs show differences based on the source. Reproduced with permission from Ref. [35]. (b) Variation in gradient shape arising from differences in equipment scale and design. The runs show differences in slope and linearity between two different systems, A and B. Note the change in linearity for system B after 1500 S. Unpublished data from Amgen. (c) Variation in retention time and peak shape caused by differences in gradient shape. Steeper gradients result in shorter elution times and sharper peaks. Unpublished data from Amgen.

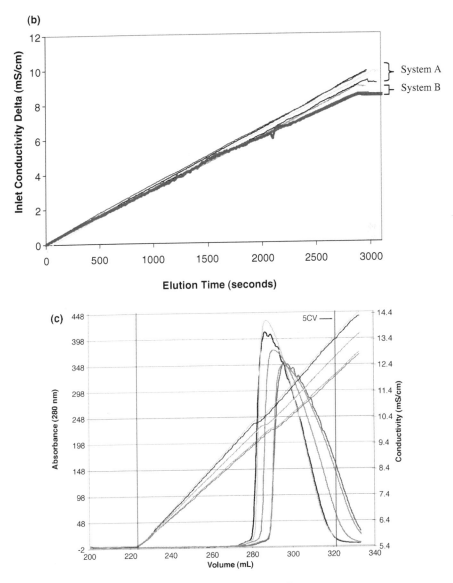

Figure 13.6. *(Continued)*.

Some of the greatest benefits from a combined PAT/QbD approach can be realized when processes are transferred from one site to the next, or one scale to the next, when there will be unavoidable changes in equipment. If the process is described and controlled in terms of process parameters and product quality attributes, and supported by well-developed models, the process becomes scale transparent and the risk associated with the transfer is understood and minimized (assuming that issues in scalability of monitoring systems are also addressed).

Full implementation of PAT occurs when product quality attributes are used directly to make control decisions during processing. UV is widely used as a general protein detection system, but it does not discriminate between product and nonproduct peaks. Analytical methods need to be very rapid to cope with the flow rates used in process chromatography. Online HPLC has been used to analyze process streams and make decisions with respect to peak cutting, but provision has to be made to provide sufficient lag between the sampling point and the valve for switching between collection and waste [37]. Feed-forward control based on analysis of the sample before loading on to the column has been proposed as a way of responding to variability of glycosyation in a highly glycosylated product [38], see Fig. 13.7. Decisions to start and stop peak collection are based on the most conservative conditions for product quality (Fig. 13.7a). If peak collection decisions were instead based on a prospective analysis of the product pool before loading on to the column, the process would deliver more consistent product quality. A more recent proposal couples analysis of process streams with predictive models and feed-forward control of sample collection [39]. In this example a predictive model was developed that would determine if collection of fraction N was appropriate based on analysis of the N-1 peak. It is worth noting that in both these cases this strategy could introduce greater variability in yields for the step in question. This may raise concerns at first glance for those who use yield as a measure of process

(a)

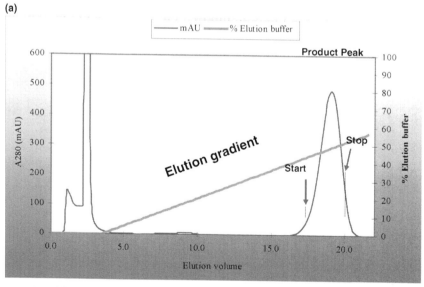

Figure 13.7. (a) Fractionation of a highly glycosylated product by a column chromatography step. Peak collection points are indicated by arrows and are set based on "worst-case" scenarios (with thanks to Dr Duane Bonam, Ref. [38]). (b) Fractionation of a highly glycosylated product by a column chromatography step. Analysis of the product pool before loading gives an indication of where the peak cut should be made to deliver consistent quality. Lower glycosylation levels would indicate delaying the stop collection point to ensure a product pool of consistent quality and permitting the collection of more product.

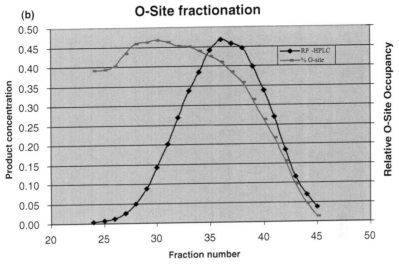

Figure 13.7. *(Continued).*

consistency, but proper consideration of the situation would address the concerns as these strategies deliver greater consistency in quality.

New sensors that would allow for real-time analysis of other product quality attributes, such as presence of host-cell proteins and aggregates, would be invaluable tools and would stimulate a rapid adoption of PAT controls. Virus removal, however, will remain the domain of traditional validation approaches for some time to come

13.5.6 Drug Product: Freeze-Drying

The final dosage form presentations of biologic (protein) medicines are typically ready-to-use injectable liquids contained in vials and syringes or reconstituted lyophilized solids in vials. The first preference from a patient convenience perspective is a presentation allowing self-administration of the medicine in one's personal space, using a once only dose and 'ready-to-use' syringe injectable supported by a delivery device. The alternative is a presentation that requires professional medical assistance as in a reconstituted solid delivered through an infusion system in a clinical institution. The selection of which of these alternatives will be the final marketed dosage form presentation is dictated usually by the injection volume, intrinsic stability of the active pharmaceutical ingredient (API) and what shelf-life stability can be achieved by the formulation excipients, dosage strength, container system, and conditions of storage. As most degradation pathways of biologics medicines are encountered in the aqueous solution state, the removal of unbound or free water from the final dosage form by lyophilization enhances the shelf life of biologic medicines and provides a common alternative to the ready-to-use liquids, which may require significant development time to discover a suitable formulation composition that provides the desired shelf life for a marketed product and rapid dissolution.

The final drug substance manufacturing process for a ready-to-use liquid drug product is substantially complete at the end of the API (active pharmaceutical ingredient) manufacture, which typically delivers the formulated material with the targeted concentration of the active ingredient in combination with all of the excipients according to the formulation recipe. The next steps in the manufacturing process for the ready-to-use liquid, if no compositional adjustments are required, fill the bulk formulation into the desired containers, vials, or syringes, according to the final presentation requirements of dosing units, followed by inspection, labeling and packaging, if no delivery device is required. The opportunities for PAT applications in the so-called "fill and finish" operations of the ready-to-use liquid drug product are mainly linked to at-line controls to monitor the metered dosages (e.g., fill weight), automated visual inspections of the final presentation for particles or cosmetic imperfections in the containers, closures and the labeling that may have a product quality impact. If the "Control Strategy" for the ready-to-use liquid incorporates the use of PAT tools to support final product release, then the at-line testing of the filled vials and syringes would require nondestructive, noninvasive tools for specific quality attributes, for example, identity by NIR. But the greatest opportunity to measure key product quality attributes by PAT tools of the ready-to-use liquid drug product as part of a control strategy is during the bulk manufacturing of the formulated dosage form. On the contrary, the manufacturing process for lyophilized solid dosage form is very amenable to the application of PAT tools (e.g., pressure, product temperature, shelf temperature, etc.) during process design, optimization, scale-up, technology transfer, and full-scale commercial manufacturing execution.

13.5.7 Application of PAT Tools for Freeze-Drying Manufacturing Processes

Designing a freeze-dried biologic formulation recipe and the freeze-drying cycle to deliver the desired final dosage form presentation in a commercial manufacturing setting is a multistep activity that requires both the formulation composition and the lyophilization cycle to be in step with each other, as the two are interdependent, that is, the formulation composition will dictate the optimum cycle design and the cycle design will influence the formulation composition. The development and optimization of the freeze-dried product start with the considerations of the liquid formulation requirements associated with chemical degradation, for example, oxidation, deamidation, isomerization. Next is physical degradation, such as unfolding, aggregation, precipitation, and solution conditions, such as pH and ionic strength. Finally, stress, due to for example, photostability, temperature, and freeze–thaw. In addition, to deliver the freeze-dried robust cake product, lyo-specific excipients are needed such as cryoprotectants or lyoprotectants, for example, glycine, mannitol, and surfactants to aid shock, freezing, and reconstitution, for example, polysorbate 20 or 80. Traditionally, to arrive at the best solution formulation recipe, the design of experimental approaches is used to establish the least number of experiments that are required to deliver formulation optimum and shortest lyo cycle.

In optimizing the freeze-drying process, the solid-state properties of the final dosage form combination of API and excipients and water must be considered in conjunction with the operating conditions of the freeze-dryer to deliver the desired product quality

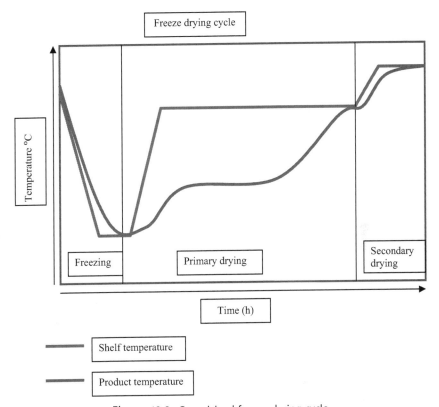

Figure 13.8. Genericized freeze-drying cycle.

attributes [40]. Key aspects of the lyophilization cycle (Fig. 13.8) must include maintaining the product temperature below its collapse temperature T_c during freeze-drying, which is the temperature at which freeze-dried cake will collapse. The primary drying phase of the process (sublimation of ice under vacuum) is the longest step, but must be optimized to be as short as possible, as this adds time and cost to the process. The secondary drying phase (desorption of water from the product under high vacuum and high product temperature) must not start until the primary drying is completed, as this may take the product temperature above T_c during primary drying.

The application of PAT tools in freeze-drying covers a very wide range of usage. One of the reasons for this is that the full operating conditions of the full-scale commercial manufacturing freeze-dryer can be duplicated easily in off-the-shelf scaled-down freeze-dryers and the physical conditions of cooling and heating, reduced pressure under vacuum, and purging with inert gases can all be executed in model systems directly linked to PAT-like tools [41].

13.5.8 Freezing

Freezing has been studied extensively to understand ice formation, crystallization, and amorphous state of API and excipients. The rate of freezing can influence ice formation, the cryo-concentration of API and excipients, which may have stability implications due to local extremes of pH or concentration [40]. Optimization of freezing has been studied using many analytical techniques. Differential scanning calorimetry (DSC) has been used routinely to determine glass transition temperature of the product solid state and to study crystallization states during annealing [42]. The modulated DSC apparatus is capable of monitoring heat flows showing exothermic and endothermic transitions over the temperature range from $-100°C$ to $0°C$.

The morphology of the solid state of the drug product and ice has been studied using cryomicroscopy. Examination of the micrographs shows the distribution of ice within the drug product and the interface between the ice and the product. Other techniques, such as X-ray diffractometry and nuclear magnetic resonance spectrometry have been used to study crystalline states of product and ice, respectively, while electrical resistance measurements have been used to study cooling rates.

13.5.9   Primary Drying and Secondary Drying

Optimization of primary and secondary drying conditions and times requires appropriate shelf temperature and vacuum to obtain the correct sublimation and water desorption rates. The sublimation end point that marks the end of primary drying has been controlled in real time using PAT tools that monitor residual water and water vapor pressure. The completion of the overall drying process (primary and secondary drying) culminates in constant water content (at the end of the targeted drying value) and a product temperature equal to the shelf temperature. Water vapor in the drying chamber, pressure of the chamber, and product temperature can be monitored conventionally in real time by sensors that provide direct end of drying information while providing feedback control to maintain constant pressure through nitrogen purging or maintaining the targeted shelf temperature during drying. Product temperature sensors are typically thermocouples or resistance temperature detectors that are placed inside the product vials. These sensors generate some heating themselves and may affect the drying process of the vials that they are monitoring. The conventional moisture sensor is typically a Pirani gauge, which detects levels of nitrogen and water vapor and provides the water vapor profile over the time course of the primary and secondary drying. The sensor for monitoring pressure in the freeze-drying chamber is typically a capacitance manometer that provides absolute pressure measurements. More sophisticated analytical instruments have been applied to the determination of moisture levels during drying. These include mass spectrometry for gas analysis, which samples the gas stream directly from the freeze-dryer in real time and produces a moisture level profile for the overall drying process [43]. Online monitoring of the sublimation end point has been approached using an electronic moisture sensor based on the determination of the partial pressure of water in the drying chamber [44, 45]. Manometric temperature measurement (MTM) is a real-time noninvasive PAT method that measures product temperature at the sublimation interface during primary drying. The MTM method generates a profile of pressure versus time data that are recorded at timed intervals during primary drying. At each measuring interval, the valve between the drying chamber and the condenser is closed for about 25 s, and the rise in chamber pressure over this time is measured. These measurements are converted by the use of a series of equations to provide data for product temperature at the sublimation interface [46], a key measure of the completion of primary drying, to determine dry product resistance to vapor flow [47], which impacts the primary drying time, and to determine heat mass transfer [48], which is equivalent to the heat transfer from the shelf and other sources in the freeze-drying chamber and is used to estimate the shelf temperature required to maintain the target product temperature. Combining all of the different data obtained by the MTM method through an expert software system has led to the development of the "Smart Freeze Dryer," which can be used to optimize the freeze-drying process in a single run [49]. Optimization of chamber pressure, target product temperature, and primary and secondary drying time can be determined by the algorithm of the expert system.

13.5.10   In-Line or Online Spectroscopic PAT Tools

Both near infrared (NIR) and Raman have been explored as real-time PAT tools for freeze-drying due to their noninvasive capabilities. Both techniques provide product quality information reflecting both physical and chemical properties of the solid drug product. Raman spectroscopy has been used in a noninvasive manner with the probe placed in the freeze-dryer but outside the product vials to collect a variety of data during the freeze-drying process [50]. Data processing using principal component analysis determined underlying factors contributing spectral variation and hence provided clear trends in the data [51]. The data provided information on the solid state of the drug product, the end points of primary and secondary drying, and physical changes occurring in the drug product during the process. The solid-state information included crystallization states and polymorph states of excipients and the crystallization of ice and the onset of sublimation. NIR has been developed for use as an in-line noninvasive tool for end-product moisture level testing of freeze-dried product in a continuous manner in a production line. The NIR spectra are recorded through the bottom of the vials in the diffuse reflection mode, with 5 scans per vial at a continuous moving rate of up to 300 vials per minute [52].

13.5.11   Scaling-Up PAT Technologies

One aspect of PAT tools, which is of some concern, is the scalability of the sensors and the measurements while going from laboratory-scale to pilot-scale and to full-scale commercial manufacturing equipment. In addition to the size difference while going from laboratory-scale to the full-scale commercial equipment, consideration has to be given to commercial manufacturing operations, which automates as many parts of the process as possible including loading of the freeze-dryer, stopper placement, and cleaning of the chamber by CIP methods another limit is the temperature ramp rates that is limited in large scale units. In addition, aseptic handling, leakage controls, and product integrity may not be compromised by the use of sensors and analyzers during full-commercial GMP production. Note that the geometry of the shelf and its loading density have an impact on the lyo speed which is a function of adjacent vials due to radiation effects. Freeze-drying PAT devices that have been developed with these considerations in mind include a cold plasma ionization device based on inductively coupled plasma–optical emission spectroscopy (ICP-OES) [53] and a tunable diode laser absorption spectroscopy (TDLAS) device [54]. The cold plasma ionization device is connected to the freeze-dryer chamber through a sampling port, and contact points at the internal parts of the freeze-dryer are able to withstand sterilization. The plasma sensor measures the ratio of water vapor to nitrogen under vacuum conditions, tracks the primary drying process, and indicates the end of primary drying. This device was reported to provide operationally equivalent measurements in tracking primary drying in four different freeze-dryers, two at pilot scale, one at intermediate scale, and a full-scale industrial freeze-dryer. The TDLAS device is installed between the freeze-drying chamber and the condenser and conducts velocity and mass flow measurements of the water vapor exiting the chamber. These concentration measurements can be used to determine the primary and secondary drying end points. Measurement of sublimation rate is useful for scaling-up drying cycles. As the TDLAS measurements are independent of the freeze-drying equipment size and capacity, shelf configurations and operational requirements, it appears very scalable and usable for both pilot-scale equipment and large-scale industrial equipment. The application of the TDLAS device in determining design space for a freeze-drying process has also been suggested [55]. Measurements of shelf temperature, product temperature, and sublimation rate would serve to define the boundaries for design space of freeze-drying processes with chamber pressure being the additional factor.

13.6 CONCLUSIONS

Pharmaceutical manufacturers and regulators are moving gradually but steadily toward QdB and PAT approaches. There is a common goal to achieve a state where we have an integrated development and quality system approach that enables reliable, flexible, and cost-effective manufacture of high-quality drug substances without extensive regulatory oversight. Although manufacturers may be most intrigued by what is meant by "without extensive regulatory oversight," it should not be forgotten that the primary goal is to ensure the general public of safe, efficacious, and cost-effective therapies supported by a blossoming pipeline.

Regulators have formed task forces and working groups and are collaborating with the industry to train investigators and reviewers and to develop new approaches to submissions. In contrast manufacturers remain apprehensive about adopting new approaches and technologies and sharing information [56]. Currently QbD and PAT concepts are mainly limited to one or two production steps, rather than whole processes, or are used in development to gather information to eliminate a step or variable [57]. There is a general agreement of the benefits of the combination of PAT and QbD approaches, but most applications are likely to be implemented in the mid- to long-term, rather than short-term.

QbD and PAT can be viewed as independent but closely linked approaches. Independently, they may at first seem to be contrasts—QbD strives for "right first time," whereas as PAT is very much concerned with continuous improvement. It is, however, in combination that they deliver their fullest potential and both support a greater concept of quality in the twenty-first century.

ACKNOWLEDGMENTS

The authors would particularly like to thank Dr. Duane Bonam of Amgen and Dr. Moheb Nasr of the FDA for permission to use figures from their presentations at IFPAC and Drs Chris Watts and Ali Afnán, both from the FDA, for reviewing this chapter.

REFERENCES

[1] Watts DC, Clark JE. Driving the future of pharmaceutical quality. *J Process Anal Technol* 2006;3(6):6–9.

[2] PAT: A Framework for Innovative Pharmaceutical Development, Manufacturing and Quality Assurance, September. 2004. http://www.fda.gov.cder/guidance/6419fnl.pdf.

[3] Pharmaceutical cGMPs for the 21st Century: A Risk-Based Approach, 2004. http://www.fda.gov/cder/gmp/gmp2004/CGMP%20report%20final04.pdf.

[4] Challenge and Opportunity on the Critical Path to New Medical Products, March, 2004. http://www.fda.gov/oc/initiatives/criticalpath/whitepaper.html.

[5] Pharmaceutical Development: Q8. http://www.ich.org/LOB/media/MEDIA1707.pdf.

[6] Pharmaceutical Development: Annex to Q8. http://www.ich.org/LOB/media/MEDIA4349.pdf.

[7] Pharmaceutical Risk Management: Q9. http://www.ich.org/LOB/media/MEDIA1957.pdf.

[8] Pharmaceutical Quality System: Q10. http://www.ich.org/LOB/media/MEDIA3917.pdf.

[9] EMEA PAT Home Page. http://www.emea.europa.eu/Inspections/PAThome.html.

[10] Reflection Paper. Chemical, Pharmaceutical and Biological Information to be Included in Dossiers When Process Analytical Technology (PAT) is Employed, 2006. http://www.emea.europa.eu/Inspections/docs/PATGuidance.pdf.

[11] ASTM E55 Committee Scope. http://www.astm.org/COMMIT/SCOPES/E55.htm.

[12] Simmons S. The future direction of ASTM E55 Committee on manufacture of pharmaceutical products. *Eur Pharm Rev* 2007;12(6):71–76.

[13] The National Technology Transfer and Advancement Act of 1995, 1995. http://www.tricare.osd.mil/jmis/download/PublicLaw104_113NationalTechnologyTransfer.pdf.

[14] Winkle H. Address to ASTM E55 Committee, 2005. http://www.astm.org/COMMIT/E55_fda.pdf.

[15] Chew R, Petko D. Commissioning and qualification: a new ASTM standard: GMP regulations. *Pharm Eng* 2007;38–50.

[16] Cohen N. Life sciences: an introduction to process analytical technology. *Control Environ Mag* 2004; available at http://www.cemag.us/articles.asp?pid=419.

[17] Macher J, Nickerson J. Pharmaceutical Industry Wastes $50 Billion a Year Due to Inefficient Manufacturing, 2006. http://msb.georgetown.edu/newsroom/news/archive/2006/10/pharmaceutical_industry_wastes_/.

[18] Cogdill R, Knight T, Anderson C, Drennan J. The financial returns on investments in process analytical technology and lean manufacturing: benchmarks and case study. *J Pharm Innov* 2007;2:38–50.

[19] Maes I. Real-time product release based on fermentation monitoring (based on NIR) and feed-back control. IFPAC–2005 19th International Forum Process Analytical Technology, Arlington, VA; 2005.

[20] Fahrner R, Blank G. Real-time monitoring of recombinant antibody breakthrough during protein A affinity chromatography. *Biotechnol Appl Biochem* 1999;29:109–112.

[21] Cooley R.Michigan Pharmaceutical Education Seminar. 2003, http://www.fda.gov/cder/OPS/cooley/sld001.htm.

[22] Lee A, Webber K, Ripple D, Tarlov M, Bruce E. Future technology needs for biomanufacturing: special session at IFPAC 2008 identifies goals and gaps IFPAC workshop. *Biopharm Int* 2008;21:12–14, also available at http://biopharminternational.findpharma.com/biopharm/News/Future-Technology-Needs-for-Biomanufacturing-Speci/ArticleStandard/Article/detail/502426?searchString=ifpac.

[23] Lee A, Webber K, Ripple D, Tarlov M, Bruce E. IFPAC 2008: A Ten Year Vision for Biotechnology Manufacturing Session. 2008. http://web.mit.edu/cbi/docs/IFPAC08_Session%20Summary.pdf.

[24] Rathore A, Johnson R, Yu O, Kirdar A, Annamalai A, Ahuja S, Ram K. Applications of multivariate data analysis in biotech processing. *Biopharm Int* 2007;20:130–134.

[25] Kirdar A, Conner J, Baclaski J, Rathore A. Application of multi-variate analysis toward biotech processes: case study of a cell-culture unit operation. *Biotechnol Prog* 2007;23:61–67.

[26] Velayudhan A, Menon M. Modeling of purification operations in biotechnology: enabling process development, optimization and scale-up. *Biotech Prog* 2007;23:68–73.

[27] Shukla A, Hubbard B, Tressel T, Guhan S, Low D. Downstream processing of monoclonal antibodies: application of platform approaches. *J Chromatogr B* 2007;848:28–39.

[28] Pujar N, Low D, O'Leary R. Antibody purification: drivers of change in Gottschalk, U (ed) Process Scale Purification of Antibodies Wiley & Sons, New Jersey; 2009.

[29] Standard Guide for Process Understanding Related to Pharmaceutical Manufacture and Control. http://www.astm.org/cgi-bin/SoftCart.exe/COMMIT/SUBCOMMIT/E5501.htm?L+memberstore+npxd3610+1206755405.

[30] ASTM Standard E2474-06. Standard Practice for Pharmaceutical Process Design Utilizing Process Analytical Technology. ASTM International, West Conshohocken, PA; 2006.

[31] Lanan M. Data management and advanced monitoring of biological processes. IFPAC Annual Meeting, Baltimore, MD; 2007.

[32] Undey C. Data and systems management towards biopharmaceutical PAT-oriented scientific manufacturing. ISPE Conference on Manufacturing Excellence; 2008.

[33] Moscariello J, Hendrickson R, Rydholm E, Hershberg R. A rapid technique to provide reproducible pressure-flow behaviour upon the scale-up of chromatography columns. 232nd American Chemical Society National Conference, San Francisco, BIOT Division. Abstract 48; 2006.

[34] Wahome J, Zhou W, Kundu A. Impact of lot-to-lot variability of cation exchange chromatography resin on process performance. *Biopharm Int* 2008; 21:48–56, also available at http://biopharminternational.findpharma.com/biopharm/Downstream+Processing/Impact-of-Lot-to-Lot-Variability-of-Cation-Exchang/ArticleStandard/Article/detail/513395.

[35] Aldington S, Bonnerjea J. Scale-up of monoclonal antibody purification processes. *J Chromatogr B* 2007;848:64–78.

[36] Larson T, Davis J, Lam H, Cacia J. Use of process data to assess chromatographic performance in production-scale protein purification columns. *Biotechnol Prog* 2003;19:485–492.

[37] Cooley R. Benefits of Process Analytical Technology (PAT) to optimization of Biotech Manufacturing IFPAC—19th International Forum Process Analytical Technology, Arlington, VA; 2005.

[38] Bonam D. Implementing PAT in Biopharmaceutical Processes: Issues and Concerns IFPAC-2005 19th International Forum Process Analytical Technology, Arlington, VA; 2005.

[39] Rathore A, Yu M, Yeboah S, Sharma A. Case study and application of process analytical technology (PAT) towards bioprocessing: use of on-line high performance liquid chromatography (HPLC) for making real-time pooling decisions for process chromatography. *Biotechnol Bioeng* 2008;100:306–316.

[40] Tang X, Pikal M. Design of freeze drying processes for pharmaceuticals: practical advice. *Pharm Res* 2004;21(2):191–200.

[41] Frinke G, Boeckem M, Steiner M. Advantages of PAT-Technologies for Cycle Development of Freeze Drying Processes, PDA Global PAT Conference, Bethesda, MD, May 22–23;2007.

[42] Milton N, Pyne A, Mishra D, Yu L. The exploration and use of thermal analytical technologies during formulation process development of freeze dried products. *Am Pharm Rev* 2003;46–51.

[43] Connelly J, Welch J. Monitor lyophilization with mass spectrometer gas analysis. *J Parenter Sci Technol* 1993;47(2):70–75.

[44] Roy M, Pikal M. Process control in freeze drying: determination of the end point of sublimation drying by an electronic moisture sensor. *J Parenter Sci Technol* 1989;43(2);60–66.

[45] Genin N, Rene F, Corrieu G. A method for on-line determination of residual water content and sublimation end-point during freeze-drying. *Chem Eng Process* 1996;35:255–263.

[46] Tang X, Nail S, Pikal M. Evaluation of manometric temperature measurement, a process analytical technology tool for freeze-drying: part I, product temperature measurement. *AAPS PharmSciTech* 2006;7(1):E1–E9.

[47] Tang X., Nail S, Pikal M. Evaluation of manometric temperature measurement, a process analytical technology tool for freeze-drying: part II, measurement of dry-layer resistance. *AAPS PharmSciTech* 2006;7(1):E1–E8.

[48] Tang X, Nail S, Pikal M. Evaluation of manometric temperature measurement, a process analytical technology tool for freeze-drying: part III, heat and mass transfer measurement. *AAPS PharmSciTech* 2006;7(1):E1–E7.

[49] Tang X, Nail S, Pikal M. Freeze-drying process design by manometric temperature measurement: design of a smart freeze-dryer. *Pharm Res* 2005;22:685–700.

[50] Romero-Torres S, Wikström H, Grant E, Taylor L. Monitoring of mannitol phase behavior during freeze-drying using non-invasive Raman spectroscopy. *PDA J Pharm Sci Technol* 2007;61:131–145.

[51] De Beer T, Allese M, Goethals F, Coppens A, Vander Heyden Y, Lopez De Diego H, Rantanen J, Verpoort F, Vervaet C, Remon J, Baeyens W. Implementation of a process analytical technology system in a freeze-drying process using Raman spectroscopy for in-line process monitoring. *Anal Chem* 2007;79:7992–8003.

[52] Sukowski L, Ulmschneider M. In-line process analytical technology based on qualitative near-infrared spectroscopy modeling. *Pharm Ind* 2005;67:830–835.

[53] Mayeresse Y, Veillon R, Sibille P, Nomine C. Freeze-drying process monitoring using a cold plasma ionization device. *J Pharm Sci Technol* 2007;61:160–174.

[54] Gieseler H, Kessler W, Finson M, Davis S, Mulhall P, Bons V, Debo D, Pikal M. Evaluation of tunable diode laser absorption spectroscopy for in-process water vapor mass flux measurements during freeze drying. *J Pharm Sci* 2007;96:1776–1793.

[55] Nail S, Searles J. Elements of quality by design in development and scale-up of freeze-dried parenterals. *Biopharm Int* 2008;21 VIS (January): 44–52.

[56] Nasr M. Strategic Implementation of PAT: FDA Perspective IFPAC 2008 Workshop, Strategic Implementation of PAT, Baltimore, MD, January 27, 2008.

[57] EMEA Report. Biologics Working Party (BWP), Process Analytical Technology (PAT) Group and Industry. Workshop on Process Analytic Technologies for Biologicals 15th March 2007. http://www.emea.europa.eu/pdfs/human/bwp/18537007en.pdf.

INDEX

Quality by Design for Biopharmaceuticals, Edited by A. S. Rathore and R. Mhatre
Copyright © 2009 John Wiley & Sons, Inc.